mw
TP08000466
KCE

Innovative Conceptual Design

Conceptual design, along with need identification and analysis, make up the initial stage of the design process. Need analysis transforms the often vague statement of a design task into a set of design requirements. Conceptual design encompasses the generation of concepts and integration into system-level solutions, leading to a relatively detailed design.

This book is devoted to the crucial initial stage of engineering design. In particular, it focuses on parameter analysis, a systematic yet flexible methodology that leads the user through the design process, helping to identify critical issues (parameters) of the design and propose configuration-specific solutions to these issues. To illustrate the principles discussed, the text presents numerous examples and a variety of real-world case studies. The emphasis throughout is on innovation. The ideas developed by the authors encourage the derivation of original solutions to new design problems or fundamental improvements to existing designs.

Innovative Conceptual Design will be a useful text for advanced undergraduate and graduate students, as well as a handy reference for practicing engineers, architects, and product development managers.

Ehud Kroll is on the faculty of the Mechanical Engineering Department at Ort Braude College in Karmiel, Israel. He was previously a professor at Texas A&M University and The University of Missouri–Columbia/Kansas City. His areas of interest include design theory and methodology; design for manufacturing, assembly, and disassembly; and automatic assembly planning.

Sridhar S. Condoor is a professor in the Department of Aerospace and Mechanical Engineering at Saint Louis University. His primary areas of interest are product design, design for manufacturability, and computer-aided design.

David G. Jansson is the principal in David G. Jansson & Associates. He was previously a professor and director of the Innovation Center at the MIT and a professor, founder, and director of the Institute for Innovation and Design in Engineering at Texas A&M University.

Innovative

Conceptual

Design

Theory and Application of Parameter Analysis

Ehud Kroll
Ort Braude College

Sridhar S. Condoor
Saint Louis University

David G. Jansson
David G. Jansson & Associates

CAMBRIDGE
UNIVERSITY PRESS

PUBLISHED BY THE PRESS SYNDICATE OF THE UNIVERSITY OF CAMBRIDGE
The Pitt Building, Trumpington Street, Cambridge, United Kingdom

CAMBRIDGE UNIVERSITY PRESS
The Edinburgh Building, Cambridge CB2 2RU, UK
40 West 20th Street, New York, NY 10011-4211, USA
10 Stamford Road, Oakleigh, VIC 3166, Australia
Ruiz de Alarcón 13, 28014, Madrid, Spain
Dock House, The Waterfront, Cape Town 8001, South Africa

http://www.cambridge.org

First published 2001

Printed in the United Kingdom at the University Press, Cambridge

Typefaces Minion 11/14 pt. and Futura   *System* QuarkXPress™ [HT]

A catalog record for this book is available from the British Library.

Library of Congress Cataloging in Publication Data

Kroll, Ehud, 1956-
   Innovative conceptual design/theory and application of parameter analysis/Ehud
Kroll, Sridhar S. Condoor, David G. Jansson.
      p.   cm.
   Includes bibliographical references and index.
   ISBN 0-521-77091-2 – ISBN 0-521-77848-4 (pb)
      1. Design, Industrial. 2. Engineering design. I. Condoor, Sridhar S., 1967. II.
Jansson, David G. III. Title.
TS171 K74 2001
620′.0042 – dc21   00-045510

ISBN 0 521 77091 2 hardback
ISBN 0 521 77848 4 paperback

To Our Families   —EK, SSC, DGJ

Contents

Contents

Preface

Over the past several years, we have taught project-based courses on product design to mechanical engineering students at several universities. We have used a variety of textbooks, but mainly our own sets of teaching notes. At the same time, we have been doing research in the area of design theory and methodology. Although most textbooks and design classes offer a decent coverage of the relevant topics, their main weakness is in handling the conceptual design stage. This is the most important phase of the design process, yet the books fall short in providing the readers with a consistent and systematic methodology to apply for different design tasks. Our lecture notes and research into *parameter analysis* as an approach to conceptual design are the basis of this book.

Innovative Conceptual Design: Theory and Application of Parameter Analysis is both an undergraduate- and graduate-level textbook for design courses in engineering programs. Practicing engineers, managers, and other product development professionals will also find the book valuable. Although our experience is with teaching the material primarily to mechanical engineering students, the applicability of the book is broad so that it may be used for teaching advanced problem solving and design in various contexts, such as other engineering disciplines and architecture.

The text focuses on the initial stages of the design process, namely, *need identification and analysis* and *conceptual design*. We define *need analysis* as the process of transforming an often vague statement of a design task into a set of *design requirements*. In this transformation process, the designer develops substantial insight into the design task, in terms of both functionality and constraints. Under *conceptual design* we encompass much more than other authors do. We include the generation of concepts, the integration of these concepts into system-level solutions, and the realization of the designs as solid and concrete working configurations. Our conceptual design phase ends with a relatively detailed design, which has been shown to satisfy the requirements of the need. Our emphasis throughout the text is on innovation. We encourage the derivation of original solutions to design problems that have not been solved before. If solutions already exist, we accentuate the need for substantial and fundamental improvements.

Most existing design books guide the reader to various search and enumeration methods in order to generate a multitude of ideas on how to solve the design task. These books refer the designer to patent and literature surveys to identify potential solutions that may be adapted to solve the task at hand. Because these approaches rely heavily on previously generated solutions, they do not foster innovation. Furthermore, they fail to provide the designer with a starting point in a totally new design situation.

Most books also offer variations of brainstorming techniques whose purpose is to organize the generation of ideas. Although these techniques provide a useful starting point for the designer, they do not help to transform an initial, rough idea into a good conceptual design. An innovative design is a synergy of a number of good ideas, not just a single good idea. Brainstorming techniques, however, do not help in identifying ideas and integrating them into a cohesive solution.

Because this book focuses on the *parameter analysis* methodology to conceptual design, it eliminates the frequent sight of design students and designers staring at a blank sheet of paper and desperately trying to create a workable design. Parameter analysis is a systematic, yet flexible, methodology that leads the user through the design process. It helps the user identify the critical issues (*parameters*) of the design and propose configuration-specific solutions to these issues. Parameter analysis fosters innovation by requiring the designer to continually incorporate new ideas into the design. It is also beneficial as a tool for educating designers to think in a systematic and creative manner.

The content of the book is ideal for project-based courses on engineering design. Such courses are already prevalent at the senior and graduate levels, and the current trend is toward earlier teaching of design methodologies. The engineering accreditation agency, ABET, and other authoritative bodies have been advocating for some time the integration of design across the entire engineering curriculum. As a result, engineering departments are struggling to find appropriate teaching materials for design. The present volume uses an approach to educating designers that is not limited to a specific level of studies. We have taught parameter analysis, the main thrust of the book, mainly at the senior and graduate levels; however, one of us (S. C.) has taught the methodology to second-year students.

Faculty who teach design at a higher level than machine elements rarely adopt a textbook author's philosophy in its entirety. Rather, they assemble their courses from articles, case studies, and miscellaneous textbook chapters. This style is common in engineering schools within the field of design. The perspective that guides our book will enable faculty either to adopt it as the sole textbook for a comprehensive and in-depth coverage of conceptual design or to use it as a resource on conceptual design if the class also deals with other topics. In both cases, the focus of our book on conceptual design will fill an existing void at both the undergraduate and graduate levels.

The parameter analysis methodology was developed by Y. T. Li and his colleagues at the Massachusetts Institute of Technology (MIT), including David G. Jansson. Dr. Jansson and his colleagues at MIT and later at Texas A&M University, including the other authors of this book, carried out further developments and refinements. In particular, for a number of years, Dr. Jansson taught a graduate class at MIT, called "Invention," with Prof. A. Douglas Carmichael. Parameter analysis was a primary part of that course. Dr. Jansson is indebted to Dr. Carmichael for the many hours of discussion, interchange of ideas, refinement of examples, and improvements in teaching methodology that were the fruit of this partnership.

Dr. Condoor and Dr. Kroll have been teaching parameter analysis at Texas A&M University, Saint Louis University, The University of Missouri–Columbia/Kansas City, Ort Braude College, and the Technion–Israel Institute of Technology. They are grateful to the many undergraduate and graduate students, academic colleagues, and professionals from industry and government who have contributed both directly and indirectly to this book.

1

Introduction

Conceptual design is the thought process of generating and implementing the fundamental ideas that characterize a product or system. This process significantly affects the product novelty, performance, robustness, development time, value, and cost. This chapter presents the framework for the entire design process, within which the upstream stages of studying the need and developing a design concept fit. It concludes with a brief "road map" to the book.

1.1 What Is Conceptual Design?

The core technical concepts developed during conceptual design fundamentally differentiate one product from other competitive products. The degree to which a new product is based on core concepts that are dissimilar from those of existing products determines its level of innovation. Innovative solutions may be viewed as designs that incorporate novel concepts and exceptional functionality. Such solutions can provide companies with a competitive advantage.

Core Technical Concepts of Computer Printers

Ink jet and laser printers serve the same function of producing an image on paper. The ink jet printer transfers small droplets of ink onto the paper, and these droplets form dots, which in turn become the image. On the other hand, the laser printer uses a laser source to trace the image on a drum. When the paper rolls over the drum, the image is transferred onto it. Although both products accomplish the function of applying ink to paper, their core technical concepts are different. This difference results from activities that took place during conceptual design.

To extend this example further, two major technologies (thermal and piezoelectric) are currently in use to inject the ink droplets in ink jet printers. The difference between the ink jet technologies, however, is smaller than that between ink jet and laser printers. Nevertheless, this difference is also the result of the conceptual design stage.

The context in which engineering design activities occur is much larger and more comprehensive than the conceptual design stage. It includes the roles of marketing, finance, planning, and overall management. Furthermore, engineering design is carried out in an integrated and concurrent manner, and requires effective communication and cooperation among the team members. The cooperation of team members is not just cross-functional but may include team members participating within a single functional area. For example, the interaction of team members within the conceptual design stage itself is often quite advantageous as team members build on or play off of the ideas of others. Note that even though designs are often done by teams, in this book we frequently refer to the creator of the product as "the designer."

The task of conceiving of new concepts and their corresponding physical configurations to meet the demands of a market need is not a simple one. In addition to the traditional engineering and scientific disciplines, this complex process involves human cognition, a field regarded as fundamentally outside of the engineer's interest or expertise. How do humans think? How do we create new ideas? What are the cognitive processes involved in a successful conceptual design activity? In spite of these questions and to the relief of most, if not all, of the readers, this book is indeed an engineering text, not a textbook on human cognition.

1.2 Parameter Analysis: A Conceptual Design Methodology

The primary focus of this book is the conceptual design process itself. We present a methodological approach to conceptual design, which we call *parameter analysis.* The development of parameter analysis involved observation of numerous successful and unsuccessful conceptual design processes, attempts to understand the relevant thought processes, and identification of what occurred as these new ideas were created and took shape. Thus, the formulation of the parameter analysis methodology mirrors successful conceptual design activity.

The term "methodology," rather than the term "method," is used purposefully to describe the parameter analysis approach. This may be a somewhat subtle philosophical distinction, but as the authors see it, the word "method" more commonly implies an orderly, step-by-step, prescriptive process with a predictable outcome. "Methodology," on the other hand, is indicative of a process that is based more on a set of general guiding principles than a series of steps. Parameter analysis does indeed describe a set of principles that are important in conceptual design and guides the designer in developing an initial idea into a design. However, through extensive experience in teaching this approach, the authors have discovered that in the early stages of learning parameter analysis, the methodical use of a formal, step-by-step process is very helpful in communicating the principles involved. Thus, this textbook uses somewhat of a "forced march" approach. When the principles are well understood, the extra effort required by such an orderly process is not always necessary.

Just as other textbooks may expand on certain elements of the larger process, this book puts a magnifying glass onto the conceptual design stage. In a sense, one can argue that conceptual design is the most difficult or at least elusive part of the engineering design process since it involves the creation of something new. (The authors recognize that often the insightful recognition of a need is the most crucial contribution to the success of a product, or that it is a clever marketing approach or stunning product aesthetics which will drive a product to the top.) At the risk of overgeneralizing, we contend that parameter analysis is profoundly different from some other approaches to conceptual design, which are more oriented to listing what is already known or searching the state-of-the-art for the best existing solution. Some of these approaches tend to cause the designer to fixate onto existing solutions, thereby hindering innovation. As the reader will see in later chapters, the parameter analysis methodology recognizes the nature of the task and focuses on ways to "unlock the unknown."

One further introductory comment about parameter analysis is in order. As the book will show, a solid understanding of the fundamentals of the physical sciences and engineering disciplines is an important component of parameter analysis. It is often true that the core concept of a clever conceptual design emerges from a thorough understanding of the fundamental physics involved. Thus, there is no substitute for continuous learning of the basics of engineering science. Learning to analyze existing products in a way that uncovers the underlying physics of a configuration is a useful and important by-product of the parameter analysis approach to conceptual design. We call this process *technology observation* and discuss it in Chapter 11. The reader will recognize the linkage between parameter analysis and technology observation in that technology observation essentially represents an effort to discover and learn from the core concepts behind the successful conceptual designs of others.

1.3 Overview of the Engineering Design Process

Numerous terms are used to refer to the process or parts of the process by which products are created, developed, and delivered to the marketplace. Some of these terms are technological innovation, product design, product development, product realization, inven-

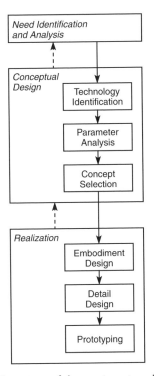

Figure 1.1 Overview of the engineering design process.

tion, and, of course, engineering design. Practitioners and academics often differ on the definitions of these terms. In addition, there are many different models of the engineering design process, but they all include the following elements in some form or another:

- A stage of identifying and analyzing a need prior to initiating conceptual design.
- A conceptual design stage to create new ideas that satisfy the need.
- Activities through which a concept is turned into an overall product or system layout.
- A stage of finalizing the design details.

A high-level view of the engineering design process is represented by the flowchart shown in Fig. 1.1.

The purpose of the upstream stage called *need identification and analysis* is to set the stage for conceptual design. As we will see in this text, the effectiveness of the conceptual design process depends in part on how well a need in the marketplace is understood (analyzed). The two major functions of need identification and analysis can be described in broad terms as:

1. Discovering what the *real* need is, and
2. Analyzing the need such that the best possible solutions to meet its requirements are not precluded by the way the need is understood or described.

In a sense, both the roles of need identification and analysis relate to removing bias from the process. Identifying the *real* need deals with recognizing and removing preconceptions of the need definition. Furthermore, ensuring that the best possible solutions are not inadvertently ruled out of the process is accomplished by analyzing the need in a way in which the influence of preconceived solutions to the problem is objectively removed from the analysis. These two influences on the whole process are inherently quite similar and suppress the natural tendencies of human problem solvers. Therefore, the overt process of ensuring objectivity at this early stage is an essential part of good engineering design practice.

The second stage in the design process is *conceptual design*. The designer will always need a "spark of ingenuity" to come up with innovative initial ideas. These ideas can come from the step we call *technology identification* or from recognizing critical issues during need analysis. One of the strengths of parameter analysis is in providing a systematic methodology to develop an initial, rough idea into a viable conceptual solution. Compared to other approaches, we observe that conceptual designs developed through parameter analysis are considerably more detailed and firm. In the conceptual design stage, the designer typically develops three or more solutions to increase the likelihood of innovation. These solutions are then evaluated against each other to determine which one will be developed further. This stage is known as *concept selection*.

The broadly defined downstream design and development activities identified as *realization* in the simple model of Fig. 1.1 include embodiment design, detail design, and prototyping. These stages may include other functions, such as proof-of-concept testing and design for manufacturing. Although this textbook does not address these downstream design activities, it is vital to recognize that the entire process is not complete until every element of the process has been accomplished.

Notice that the model recognizes that the interfaces between conceptual design and realization, and between need identification and analysis and conceptual design, are bidirectional. That is, it is possible that a conceptual design will be "sent back to the drawing board" or rejected altogether if the results of the downstream efforts reveal issues overlooked during the need analysis and conceptual design stages. In other words, a bad conceptual design cannot become a good design regardless of the quality of the realization effort and may therefore call for additional work. This can cause lengthy delays and increased product development costs, and will eventually have an adverse effect on the success of the product.

1.4 Structure of the Book

This book is heavily dependent on examples to elucidate the methodology presented, and the examples are carefully detailed in a way that clarifies the thought processes undertaken. The student reader is encouraged to study the examples and case studies fully rather than merely read them through rapidly. Furthermore, the student is encouraged to enhance his or her understanding of the methodology by detailing each part of the thinking process in the formal way that is presented in the book. As already mentioned, when the parameter analysis methodology becomes more familiar, the formalism of this approach will be less important. However, successful, high-quality conceptual design comes about through practice and hard work, as well as inspiration. It is often argued that clever, creative conceptual design skills are a gift, not something that can be acquired by learning. However, the authors have seen first-hand the value of the "coaching method" called parameter analysis. Just as an incredibly gifted athlete must practice (with coaching!) to reach the highest level of competitiveness, so, too, can designers benefit from a methodology that sharpens their ability to think in ways that foster creative solutions.

The authors have considerable experience in teaching parameter analysis. This approach to conceptual design has been tested in both graduate and undergraduate design subjects and in numerous industry short courses and projects. The combination of "theoretical" discussion of the principles of parameter analysis and the use of numerous examples and exercises has proved to be an effective way to learn this material.

Chapter 2 introduces a methodology for identifying and analyzing needs as the preparatory step to conceptual design. Chapter 3 presents a detailed example of need analysis to illustrate the formalisms of Chapter 2. While Chapter 4 introduces the parameter analysis methodology, Chapter 5 offers an expanded discussion and provides many guidelines for implementing the methodology. Chapters 6 and 7 are case studies of applying parameter analysis to relatively simple sensor design tasks. Development of alternative conceptual solutions to meet an industrial need is demonstrated in Chapter 8. Chapter 9 shows how parameter analysis was applied to the design of a machine for a novel extrusion process. Chapter 10 contains a comprehensive need identification and analysis, together with the subsequent conceptual design for entry in an engineering students competition. Technology observation is presented in Chapter 11 as an approach that will sharpen the skills of designers by understanding the concepts behind the designs of others. The book concludes with Chapter 12, which provides direction on how to proceed downstream in the design process, and discusses some of the cognitive aspects of parameter analysis.

2

Need Identification and Analysis

Every design process begins by identifying and analyzing the need, and defining it in terms of the design requirements. Clarifying and quantifying the design task helps the designer to gain crucial insights that will facilitate the creation of innovative products. This chapter provides a methodology for carrying out these initial steps. The methodology first calls for identifying the real need in qualitative and solution-independent terms. Next, the overall need is studied in light of five basic categories of functions and constraints: *performance, value, size, safety,* and *special.* These functions and constraints should be fulfilled and satisfied by any design solution that will emerge later in the design process, yet any premature specification of design solutions is carefully avoided during this stage. The need analysis is summarized at the end of the process as a consistent and quantitative set of design requirements.

2.1 The Importance of Need Identification

Design tasks arise from a variety of sources, sometimes referred to as "customers." Because these sources are often nontechnical, the tasks may not be defined in engineering terms. Furthermore, initial task statements sometimes describe *perceived* problems as opposed to *real* needs. The natural tendency of novice designers is to start thinking of solutions that satisfy the perceived problem before gaining full understanding of what should really be designed. This tendency, coupled with a lack of proper definition of the task, may mislead the designer and waste his or her time and effort in solving the wrong problem.

Successful identification and definition of a market need is a prerequisite in developing innovative products. Many companies actively identify new market needs and technologies through market surveys, product benchmarking, technology observation, and forecasting. These activities enable companies to identify immediate, interim, and long-term market needs, as well as current and emerging technologies. Later, during the design process, the designers will focus their efforts on establishing synergy between market needs and existing and new technologies.

An incorrect need definition focuses attention on solution-specific issues, and so the designer will be fixated on developing an apparent solution into a product. In the process, many superior solutions that fit the overall design need better may be overlooked. Competition can leapfrog by better defining the need and conceiving more innovative solutions that match the current technologies with the real need. The importance of recognizing and understanding market needs and matching new technologies with the changing needs of society is illustrated by the following example of the evolution of typewriters.

Market Needs: Insights from the Evolution of Typewriters

The typewriter, a great American invention, shaped our society and the mode of communication for over a century. The origin of typewriters can be traced back to 1868, when C. L. Sholes and several of his associates created the first typewriter prototype. E. Remington and Sons began selling a refined version of this machine in 1874. This typewriter was popularly known as the "blind writer" because the paper was hidden from the typist's view. The character set consisted of uppercase letters only, but an important feature was the arrangement of letters on its keyboard, known as "QWERTY" (the first six letters at the left end of the top row). This arrangement, which helped to avoid jamming of the keys at high speeds by separating the letters that are often typed in succession into opposite sides of the keyboard, persisted over the years and is even used in present-day computer keyboards.

To improve the usefulness of typewriters, the Smith Premier Typewriter Company introduced the Smith Premier, which used a full keyboard with separate keys for upper and lowercase letters. Remington, on the other hand, came up with the innovative Remington Model 2, which used the shift key to type both upper and lowercase letters using a single character set on the keyboard and dual-faced typebars. This keyboard represented an ergonomic improvement, reducing the mechanical movement of the hand. Even though the demand for typewriters was virtually nonexistent at the time, the few typewriter manufacturers firmly believed in the great potential of the product.

Let us now step back and examine the possible needs that typewriters could satisfy in the later part of the nineteenth century. Typewriters could satisfy a need for "clear and legible documents"; however, typewritten documents were not considered socially acceptable then, so this need did not exist. On the other hand, with the increasing speed of communication at that time, a need existed for "high-speed transcription." Typewriters could potentially satisfy this need, but very few people knew how to type. There were neither books nor instructions to teach typing. As a result, typing speeds were much lower than manual writing, and it did not make sense for a customer to buy a typewriter when he or she could write faster. Although the inventors believed that typewriters could help in high-speed transcription, they did not understand the big picture in which the typewriter, and the training to use it, together formed the complete solution. The initial market failure of typewriters can be attributed to this lack of recognizing the real market need.

The breakthrough for typewriters came in 1881, when the Young Women's Christian Association (YWCA) thought of typing as a career for women and started offering typing courses. The business community immediately recognized the need for typists, which also made way for women to enter the workforce in large numbers. To cope with the demand for training courses, several institutions opened across the country. The training, along with the typewriters themselves, now fulfilled the need for high-speed transcription. It sparked a great increase in the sales volume and in the competition as well. But while the demand for typewriters and typists was increasing, users still had to cope with the problem of "blind writers." In 1895, recognizing the need for "seeing the document while typing," the Underwood Company introduced the revolutionary Underwood No. 5 typewriter. This was the first "writing-in-sight" or "visible-writing" typewriter.

The next innovation in typewriter design addressed the need for "portability and noise-free performance." Electric technology was incorporated into typewriters in 1920, creating the electric typewriter. It had several advantages over the manual typewriter. The electrical energy was used to actuate the keys instead of finger pressure and to do the line spacing and carriage return functions. Consequently, the performance—typing speed and quality of print—was greatly improved. Recognizing these benefits, all typewriter companies moved quickly toward electric typewriters.

In 1934 Dvorak, one of the founding fathers of industrial engineering, developed a new keyboard arrangement for typewriters. The arrangement, shown in Fig. 2.1, was based on twelve years of studying languages and the physiology of the hand. In this keyboard, 70% of all English words can be typed using the middle-row keys as opposed to 30% in the QWERTY arrangement. Also, the most common consonants are placed on the right side of the keyboard, whereas the most frequently used vowels are on the left. As a result, the typist continuously uses both hands in every word. This arrangement is easier to learn and use, improves typing speed by about 10%, increases accuracy, and reduces fatigue. Despite its tremendous advantages, marketing this keyboard by Smith-Corona failed. While designing the keyboard, Dvorak overlooked an important implicit constraint: the entrenchment of the existing technology brought about by millions of people being proficient at using QWERTY keyboards. A change

Figure 2.1 Dvorak's keyboard arrangement which separates frequently used consonants and vowels into the right and left sides of the keyboard to improve typing efficiency.

would require retraining people and buying new typewriters. This constraint made the revolutionary keyboard impractical.

Even though the QWERTY arrangement reduced the tendency of the printing arms to collide, all typewriters, both electric and manual, were jamming at higher speeds. Recognizing this limitation of mechanical levers, IBM introduced a revolutionary concept, dubbed "Selectric," in 1961. This machine used a ball with letters embossed on it. The ball was mounted on a carriage, which continuously moved in the horizontal direction during typing as opposed to moving the paper relative to stationary keys. In addition to obviating the problem of jamming the levers, the new concept reduced the overall size by eliminating the carriage motion. It also reduced the moving mass, which in turn improved typing speed. The ball could also be changed easily to accommodate different languages and fonts, or replaced when worn.

The electronic era found the typewriter companies trying to cope with the new technology by introducing electronic typewriters. These devices allowed the typist to check and revise each line before printing to eliminate mistakes and the need for corrections. Electronic typewriters were further improved by incorporating a monitor to allow checking whole documents. However, these changes could not keep up with the consumers' need and appreciation of the greater flexibility offered by personal computers.

Recognizing the coming end of the typewriter era in the United States, the Brother Company began increasing sales in foreign countries. By producing different-language typewriters, they successfully exploited the residual value of the typewriter technology while switching to the next-generation products. The company diversified into printers, fax machines, word processors, and electronic stationery such as labeling systems and stamp creators. The Smith-Corona Corporation, on the other hand, was unable to make the leap from electrics to electronics and filed for bankruptcy on July 5, 1995. Interestingly, on the same day, Bill Gates, the founder and chairman of Microsoft Corp., was announced to be the richest person in the world.

2.2 Need Identification in Practice

Design tasks may originate from sources external or internal to the organization. Externally, the market is the driving force for developing new products. Design tasks may also originate internally from within the organization. For example, while designing a large system, the system integration team may define a subsystem as the design task for another team. Design tasks often originate from the manufacturing department as a need for improvement in the product manufacturability or a need for new jigs and fixtures. Regardless of the source of the task, the design process is always initiated by a task statement.

Many design task statements tend to be highly configurational. In other words, they identify the required end product rather than emphasize the need. "Design a shaft," "design an encoder," and "design a ship canal" are some typical statements. They quickly evoke existing configurations in our minds. Our natural inclination is to take these general configurations and make them more concrete. However, recalling existing configurations tends to fixate the designer onto similar and common solutions and to prevent the conception of highly innovative and effective products.

Companies often make a similar mistake by defining their core business in terms of products and not the *need* satisfied by those products. As the products become obsolete, the companies, too, face extinction, like most typewriter manufacturers. Successful companies redirect themselves with changing technologies. The next example illustrates how a new market need can emerge and emphasizes the importance of viewing the core business in terms of customer needs.

Zero Emission Vehicles

Automobiles contribute well over half of the emissions that damage the ozone layer and a significant percentage of particulate matter in the atmosphere. To address the problem of air pollution, the California Air Resources Board adopted the Low Emission and Clean Vehicle Regulation in 1990. According to this regulation, 10% of the cars sold in the year 2003 should be Zero Emission Vehicles (ZEVs). Other states and countries are following this lead and adopting strict federal standards for vehicle emissions. This regulation created a market need for ZEVs and spurred an interest in the development of electric vehicles. Several technological leaps have been made, including the development of advanced batteries that do not use lead and the commercialization of fuel cells and other alternative technologies.

This regulation had a profound impact on the oil industry. Petrochemical companies are radically restructuring their core business. Whereas earlier they viewed their core business as "supplying petrochemicals," now they redefine it as "meeting energy needs." This new definition allows the companies to expand into new markets, such as electric and hydrogen recharge stations, and might be crucial for their long-term survival.

To suppress the tendency of being configuration- or product-oriented, the designer must always begin by determining what the real need—the benefit sought by the customer—is. The need for a shaft, for example, can be defined as "transmitting torque." Leading companies spend considerable effort in defining their core business in terms of customer needs. Kodak, for instance, defines its core business as "imaging," which includes digital photography. Such definition facilitates the transition into new technologies in a systematic manner. The following two examples expand on the advantages of a good need definition.

NASA's Mars Sample Return Mission

One possible initial task statement for NASA's Mars sample return mission may be "to design a space mission to Mars that will deploy a vehicle for collecting rock samples at several locations and bring the rocks back to Earth." From this statement, one can very easily "jump" into initiating a design process of a spaceship that will travel to Mars, land on the surface, release a ground vehicle for collecting samples, take off from Mars with the samples, and travel back to Earth. However, the danger in designing this mission according to the highly configuration-specific initial task statement is that more innovative and economical options may be ignored.

Scientists have recently identified a rock found in Antarctica as a Martian rock blown into space as a result of a meteorite impact on Mars. So perhaps we can send a powerful bomb to explode on Mars, blow pieces of it into space, and wait for some of them to arrive here. We can even attempt to capture the rocks in Earth orbit by the space shuttle. Alternatively, we could send a spaceship to Mars orbit, drop bombs and collect rock fragments while in orbit, and travel back to Earth. Or, if we insist on landing a sample-collecting device on Mars, perhaps it does not necessarily have to be ground based. One Russian proposal for a similar mission consisted of a balloon, which inflates and deflates automatically due to the day–night temperature differences on Mars and is carried around by Martian winds. Another option could be to analyze the soil on Mars and only send the results to Earth.

Such innovative solutions would have been considered had the need been stated as "to obtain Martian soil samples for analysis" or "to analyze Martian soil samples." Thus, the way the need is stated may affect the degree of innovation in the design solutions.

Ship Canals

Let us examine another task statement, that of "design a ship canal to connect two bodies of water." The first thought that this statement brings to mind is digging a simple "moat," which is a possible solution when the water level of the two seas is the same. An example is the Suez Canal, which connects the Mediterranean with the Gulf of Suez at the head of the Red Sea, as shown in Fig. 2.2a. Because it is a sea-level canal, the elevation of a ship does not change during its journey. The Suez Canal remains the world's longest canal and can still be widened or deepened relatively easily to accommodate bigger ships.

After building the Suez Canal, the French tried to build the Panama Canal to connect the Caribbean Sea with the Pacific Ocean. Their attempts to dig a sea-level canal lasted nine years but ended in failure. This led the Americans to build the Panama Canal with locks that raise and lower the waterway, as shown in Fig. 2.2b. When a ship enters a lock in the canal, the gates are closed behind it and the water level is raised or lowered,

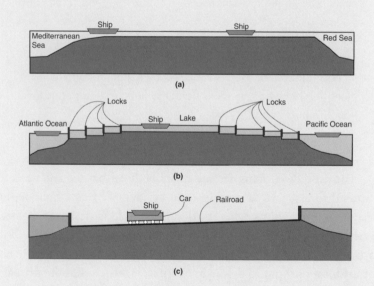

Figure 2.2 Longitudinal profiles of (a) the Suez Canal, (b) the Panama Canal, and (c) "canal on wheels" (not to scale).

thereby changing the elevation of the ship. Then the next set of locks ahead of the ship opens to allow the ship to proceed. In this way, the locks eliminate the need to maintain constant water level in the canal. Thus, a better definition of the original design task might have been "to connect the two bodies of water." Note that the last statement is less configurational (no "canal") and more functional ("connect").

If we investigate the need further—by asking, "do we really want to connect the two bodies of water?"—then the answer reveals that the real need is "to transport ships across a land-mass in a cost- and time-effective manner." The designers at Ronquiéres in Belgium utilized this insight in constructing a system for transporting ships across a 1-mile stretch of land with a 220-ft rise between Brussels and Charleroi. The gradient for this system is much larger than the Panama Canal's, which elevates ships by 85 ft over about 2 miles. Because of the steep gradient, it was difficult to implement a canal system that uses locks to raise the waterway. The designers conceived a novel system that uses 300×40-ft water-filled railroad cars to transport ships up and down the five-degree slope as shown in Fig. 2.2c. Had the designers followed the natural instinct of designing a ship canal that connects two bodies of water *with water,* they would have ignored innovative solutions such as the "canal on wheels." Therefore, before making any configurational commitment, it is necessary to understand the need in terms of its functional requirements.

The preceding examples and insights show that the *real* need, as opposed to a *perceived* need, should be identified as a first step in design. In other words, what the designer is told to design is not necessarily what the customer really needs, but rather what the customer *thinks* that he or she needs. The designer should therefore use the customer's input to define in qualitative terms what the primary purpose or goal of the product is. This definition constitutes the *need statement.*

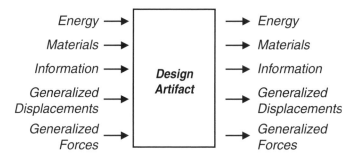

Figure 2.3 Black-box model of a design artifact with inputs and outputs.

To develop a good need statement, it is often useful to view the product-to-be as a black box with inputs and outputs. The purpose of the design artifact is to transform the inputs into desired outputs. Typically, inputs and outputs for engineering products are concerned with energy, material, and information. In mechanical engineering, it is often convenient to view some inputs and outputs in terms of generalized forces (including moments) and generalized displacements (including rotations) rather than energy. Figure 2.3 shows the representation of a design artifact as a black box that transforms inputs into outputs. The designer should identify all the inputs and outputs and define the transformation that the product must perform. In some design tasks, the required output is clear, and the designer is free to select the inputs. In such situations, the need statement should reflect only the outputs. Table 2.1 shows the need statements for a few common devices, generated by applying the above procedure. These statements in turn can lead to alternative solutions and thereby foster innovation.

2.3   The Need Analysis Methodology

Once the real need has been identified, considerable effort should be put into thoroughly understanding and precisely defining this need. On an abstract level, we can think of the collection of all possible, still-unknown solutions to the design task as forming an imaginary "solution space." As shown in Fig. 2.4, this space is enclosed by

Table 2.1 Need statements for some products based on the black-box model.

| Product | Black Box Description | | Need Statement | Alternative Solution |
	Inputs	Outputs		
Slider-crank linkage	Linear motion	Rotational motion	Convert linear motion into rotational motion.	Rack and pinion
Tachometer	Rotational motion	Speed information	Measure the rotational speed of a shaft.	Encoder + Clock
Rolling-element bearing	Rotational motion; forces	Rotational motion; forces	Transfer radial and axial forces while allowing relative motion.	Magnetic bearing

boundaries that are the constraints of the task. During the need analysis stage, the designer outlines the solution space with precise and quantitative boundaries. Although this process can easily be done by considering a specific solution, the result will be biased and narrow in scope, eliminating many potentially good designs and hampering innovation. The objective during need analysis is to maximize the size of the solution space, so the designer must refrain from inventing a solution while defining the boundaries. Indeed, thinking in solution-independent and generic terms is one of the most challenging aspects of the entire design process. The benefits of such abstract thinking include a larger solution space, better potential for innovation, and greater flexibility.

Need analysis is concerned with two main aspects of the design: *functions* and *constraints*. The purpose, or main function, of the product to be designed has already been recognized during the need identification stage. Now, however, this function needs to be further studied, refined, quantified, and perhaps broken down into subfunctions. The brief and qualitative task statement will be turned into more elaborate and quantitative descriptions of what particular characteristics the product is required to have. A complex design may require division of the overall task into smaller, more manageable subsystems that can be assigned to several designers or design teams.

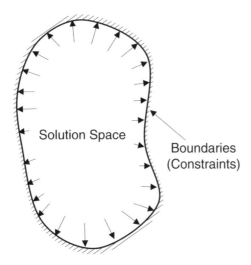

Figure 2.4 Schematic of a solution space bounded by constraints. The task of need analysis is to maximize the size of the solution space.

In an ideal situation, the designer is concerned with satisfying functions only, so the solution space is infinitely large or unbounded. A high degree of innovation is possible because any solution is acceptable. In the real world, however, the solution space is bounded by constraints (see Fig. 2.4). The constraints limit the ways in which functions can be realized and thus reduce the size of the solution space. For example, the need "to archive ideas" can be satisfied by several solutions such as using pens, pencils, audio and video recorders, and computers. Posing a constraint, for instance, adding the phrase "on paper" to the need, eliminates several potential solutions. Adding the word "erasable" as another constraint crosses out most solutions other than pencils. In this way, each additional constraint can drastically reduce the solution space, even to the point where no solution is possible.

Market surveys provide a large amount of information about customer wishes and preferences. Before turning these preferences into constraints, the designer should evaluate each item to check whether it is real or fictitious. A real constraint is one for which its violation renders the product nonprofitable or nonviable. The list of constraints should include all the real constraints because these eliminate impractical solutions and focus the design effort on feasible solutions only. Fictitious constraints, on the other hand, should be ignored to increase the solution space and foster innovation.

Constraints can be generated by examining the original design task statement, relevant regulations, and previous design requirements for similar tasks. They can be classified into *explicit* and *implicit* constraints. Explicit constraints are those listed in the initial task statement or directly derived from it and are therefore easy to identify. Implicit constraints, on the other hand, are those that must be generated during the need analysis stage and are much more important and difficult to recognize. Explicit constraints are typically listed first, and all the product designers are made aware of them to eliminate the possibility of creating a product that violates these constraints. Any changes to the explicit constraints require the customer's approval. In contrast, implicit constraints are found by studying the life-cycle environment and doing order-of-magnitude calculations. If an implicit constraint, generated during need analysis, is found to cause a problem later in the design process, the designers may reexamine and revise it.

Need analysis is a discovery process wherein designers find critical issues and gain crucial insights into the design task. To facilitate the discovery process, need analysis is divided into five general categories as shown in Fig. 2.5. These categories allow the designer to focus on a few issues at a time while continuously acquiring more understanding of the task. Studying trade-offs, such as performance versus value, is a very important activity that may even lead to changing the original need definition. In trade-off studies, the designer identifies potential inconsistencies and tries to resolve them. This adds to the designer's understanding of the task, prevents conflicts between designers working on different aspects of the design in later stages of the design process, and assures that the final result would have a good balance between the often conflicting requirements.

Need analysis is not necessarily a structured and sequential process, and designers often backtrack, both within the need analysis process, and from later in the design process back to need analysis. The designer may gain new and important insights during the conceptual design stage, which would force a revision of the need analysis and the design requirements. However, although designers should regularly examine and update the design requirements, the goal of need analysis is to minimize such time-consuming iterations. This can be achieved by investing the adequate effort in a good

```
1. Performance
2. Value
3. Size
4. Safety
5. Special
```

Figure 2.5 The five general categories of functions and constraints.

need analysis and considering downstream design issues before starting with conceptual design. Keeping the number of modifications to the design requirements to a minimum usually improves the overall time-to-market of a new product.

The recommended practice for doing need analysis is first to list all the questions and issues that should be considered, preferably classified under the five general categories of Fig. 2.5. Next, the necessary research and investigations are carried out, and the results used to answer the questions and fill the relevant issues with information. The next five sections discuss the general categories in more detail, followed by a description of how to turn the need analysis into design requirements.

2.4 Performance Considerations in Need Analysis

The primary concern when analyzing a need is to identify everything that the product must do and sometimes also how it should be done. Performance deals with functions, which are the actions or activities that a product should perform. An important characteristic of functions is that they can be decomposed into smaller or lower-level functions. This decomposition helps to break the overall task into manageable chunks, allowing several design teams to work independently on different functions and thereby greatly reducing the time-to-market. However, a bad or incorrect functional decomposition may deprive the teams of their independent decision-making ability. In such a situation, the teams are faced with conflicts on a regular basis and need to repeatedly negotiate with each other. This drastically increases the product development time. In large design tasks, the common practice is to have a system integration team that is in charge of high-level functional decomposition, interface definition, and conflict resolution. The following example illustrates some of the complexities faced by integration teams.

Undersea Remotely Operated Vehicles

Remotely operated vehicles (ROVs) are used extensively in deep-water applications in which maintaining a human diver for any length of time is cost-prohibitive. A typical ROV (schematic shown in Fig. 2.6) consists of a robotic arm that performs the desired operations and a vehicle that maneuvers the arm. The vehicle is controlled and powered from the surface through a tether. The design of the tether is an interesting challenge because the teams working on other subsystems raise conflicting demands on the tether designers.

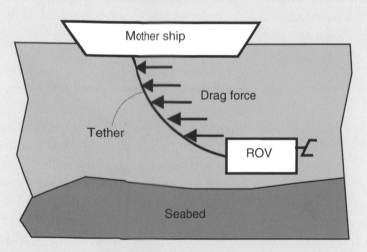

Figure 2.6 Remotely operated vehicle system.

The tether is the lifeline of the ROV: it transfers energy for navigation and control signals to the vehicle, and it transmits visual information to the mother ship. This information and energy transfer is crucial for successful operation. The tether is neutrally buoyant to reduce the gravitational forces on the ROV. However, the drag force acting on the tether due to ocean currents could be significant, requiring the ROV to continuously resist it. As a result, the tether is not only a lifeline, but also a burden on the ROV's limited resources. These considerations result in three functions in the tether design task: transfer energy, transfer information, and reduce drag force.

To satisfy these functions, the tether design team can modify several key variables, for example, the cable diameter and length. A short cable with a large diameter is preferred for effective power transmission because it reduces power losses and voltage drop. Such a cable also increases the communication bandwidth and rate. The initial task requirements usually establish a lower bound for the length, leaving the designer with more freedom to change the diameter. However, an increase in diameter affects the drag force, which in turn increases the power requirement.

Different subsystem teams, such as communication, power, and navigation, often place unreasonable demands on tether designers to increase the rates of energy and information transfer while reducing the drag force. It is the duty of the system integration team to decide on the acceptable magnitude of the energy transfer rate, information transfer rate, and drag force in the initial stages of the design process. The skill of the integration team lies in anticipating potential conflicts ahead of time and in finding suitable trade-offs that optimize the overall performance of the ROV.

Functions can be classified into *primary* and *secondary*. Primary functions are those functions that must be satisfied to fulfill the need. Therefore, all design solutions must address the primary functions. In contrast, secondary functions are not directly related to the need and tend to be solution-specific. Typical issues addressed by secondary functions are product inefficiency and protection from undesirable inputs. The following gearbox example elaborates on the differences between primary and secondary functions.

The primary function of any gearbox is to provide a constant speed ratio between input and output. Any solution, such as spur or helical gear pairs, must therefore satisfy this function. Helical gears, however, always generate a thrust load. If a successful implementation using helical gears is sought, then the designer must address the additional function—"resist the thrust loads." This function is solution-specific because another solution, using spur gears, does not require it. Such functions are known as secondary functions. "Dissipate the heat generated in the bearings" is another secondary function because it does not directly involve fulfillment of the overall need.

Designers take an active approach when addressing functions: they incorporate certain features, components, or assemblies to handle each function. As the number of functions increases, both the magnitude of the required design effort and the product cost also increase. The designer must reduce the number of functions to positively affect these two issues. Designers have very little leverage when it comes to primary functions because these are required to satisfy the overall need. Secondary functions, however, can often be eliminated. Because secondary functions often address issues such as robustness, the designer must trade cost against the product improvements related to secondary functions. The designer should therefore spend time at the beginning and decide what functions are more important. The following example of the difference between floppy disks illustrates the importance of identifying the significant secondary functions.

Computer Floppy Disks

Let us examine the need satisfied by floppy disks. At an abstract level, both 5.25″ and 3.5″ disks are aimed at fulfilling the same primary function: to store information. Both disks satisfy this need using the same concept. However, a drastic usage shift away from 5.25″ disks and toward 3.5″ disks took place in the early

1990s for several reasons. These reasons will help us to understand the importance of identifying the correct set of functions.

From a cursory examination, it is obvious that a 3.5" disk stores significantly more information in less space. The 3.5" disks benefited from advances in technology and exhibit superior portability. When 5.25" disks were introduced, their information storage density corresponded with the state-of-the-art technology of that time. Consequently, these disks were rather large, and portability became a major concern. A detailed examination reveals that 3.5" disks satisfy additional functions; for example, they can be slipped into a shirt pocket without worrying about possible damage to the information. In other words, 3.5" disks address the additional function "provide information safety," and this function is realized in the form of a rigid cover and a protective slide shutter. The protective shutter prevents accidental damage to the magnetic medium when not in use.

The 5.25" disk is perfectly square (as shown in Fig. 2.7), and as a result, it can be inserted into the drive in eight different ways. This is a minor nuisance for the user whose attention is required during disk insertion. In contrast, 3.5" disks (also shown in Fig. 2.7) are rectangular, measuring $3.5'' \times 3.7''$, and the slot in the disk drive measures 3.6". Because the disk cannot be inserted along the 3.7" side, four possible orientations are eliminated. Two more erroneous orientations are taken

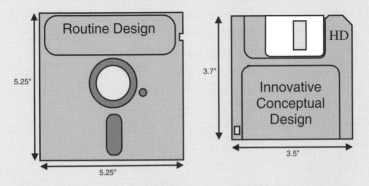

Figure 2.7 Shape and dimensions of 5.25" and 3.5" floppy disks.

away by the tactile feel provided by the metal shutter, and the final position for correct insertion is determined by the chamfer at the right-hand corner. This design therefore addresses the secondary function: "eliminate wrong insertion of the disk."

The designers of 5.25" disks addressed the function of information storage and paid very little attention to secondary functions. Although the 3.5" disk designers could benefit from their predecessors' mistakes, a good need analysis could additionally have revealed these important secondary functions to the 5.25" disk designers. These designers could have created a more robust disk easily, although it would probably have been more expensive. The designer of a successful product must therefore identify a good set of functions from the numerous demands and wishes of the customer.

The product development time may be long when new technology is involved. In such cases, it becomes very important to accurately estimate the future performance requirements. Companies often perform detailed technology forecasting and market studies to identify trends. For instance, the oil industry found that the trend in fuel usage has been to reduce the carbon-to-hydrogen ratio. Fuel cells, which combine hydrogen with oxygen to produce electricity and water, use no carbon and therefore will have the lowest rating on this scale. This trend helps companies to determine in which technologies to invest. For example, DaimlerChrysler planned to invest about $1.4 billion in five years before the release of its first commercial fuel cell-powered car in 2004.

In the semiconductor industry, Moore's law predicts the increase in the number of transistors in a chip: the number of transistors in a unit volume doubles every eighteen months. The law is based on observations by Gordon Moore, cofounder of Intel Corporation, and it helps predict the desired performance for future computers and electronic products.

Although performance considerations involve mainly functions, performance-related constraints should also be recognized and precisely defined. One of the most relevant types of constraint is time-related. Temporal constraints include the relative sequencing of different events and transfer rates. For example, the required time for deploying an automotive airbag is very short, of the order of one-twentieth of a second. This constraint requires that an explosive release of gases be used. The required deployment time for a side-impact airbag is even shorter, making the design task very difficult. Typically, temporal constraints can be defined by recognizing which events have to be performed before others and organizing them in a flowchart.

Real-life designs are first and foremost a compromise among various performance objectives. The disk brakes mounted on the wheels of a passenger airliner, for example, do not contribute at all to in-flight performance such as payload, speed, and range. However, maximum braking performance is expected on the very rare occasion of an aborted takeoff: a fully loaded plane accelerating on the runway when the pilot decides to stop because of an engine problem or an obstacle ahead. Under these severe conditions, applying the brakes to stop the aircraft would cause so much kinetic energy to be dissipated as heat that the brakes would practically melt. However, this design is very efficient: the plane does not have to carry the extra weight of more durable brakes at the expense of decreased payload, and the brakes can be replaced after an aborted takeoff, which happens very rarely.

Several calculations are often done to identify possible conflicts between the extent to which various functions can be satisfied. These trade-off studies are important because they can provide an order-of-magnitude sense of the difficulties in realizing the design together with significant insights into the task. The calculations may begin with energy and mass balances, and require making several approximations. These calculations should provide information about the characteristics of ideal solutions and the ideal performance values. Then, the designer can strive to incorporate these characteristics into the design solution and achieve the ideal performance during the design process.

2.5 Value Considerations in Need Analysis

Addressing the issue of value means much more than just cost because sometimes value cannot be directly measured in monetary terms. For example, when the design frees humans from dangerous or repetitive jobs, it makes a social contribution. A space exploration mission may make scientifically invaluable contributions, and an improvement to automotive emission control systems may have environmental value. Even the Eiffel Tower has value as a great engineering marvel, although it may not have much monetary value.

Although cost is usually associated with a specific solution, an analysis of the value of the new design can be done before its conception. Value is strongly related to the original need, the function of the product, and the level of performance attempted. There is usually a direct mapping between value and performance, as shown schematically in Fig. 2.8. A low level of performance is associated with low value, and high value is perceived as commanding superior performance. Consider an expensive luxury car that costs much more than what one would pay for a more basic vehicle. The luxury car probably represents good value to some customers, who seek the high performance levels reflected in such factors as speed, acceleration, handling, safety, comfort, reliability, and more.

In addition to requiring that the design provides certain "valuable" functions, value-related constraints should also be investigated. Marketing is a key element in the overall product development strategy. To better meet the needs of customers, markets are segmented based on economic and social considerations. The target market segment dictates the product cost and the anticipated sales volume. Assuming that the market is rational (which may not always be true), increasing the price decreases the sales and vice versa. During need analysis, attention should be paid to the target product cost and the sales volume.

From the manufacturer's perspective, capital cost, payback period, and return-on-investment are some of the key constraints related to value. The customer, on the other hand, is interested in a low product cost and a high product value. Often, the interests of customers and manufacturers are in conflict with one another. An attempt should be made to strike a balance between these competing demands.

The price of a product refers to the amount that a customer would pay to buy it. By investing in this purchase, the customer either generates revenue or saves expenditure. This monetary bene-

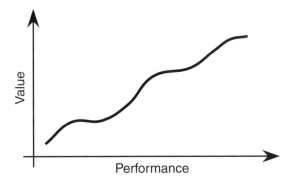

Figure 2.8 Schematic of value–performance correlation.

fit to the buyer is the value of the product. A successful product should exhibit a good match between price and value; that is, customers would usually be willing to pay more to get better value (and thereby performance).

The calculations to determine the performance–value relationships are very important. For instance, in designing time-saving machines, these calculations may redefine the task and impose new constraints on the product cost.

Making a profit is often the key motivation for developing a new product. Profit depends on the product manufacturing cost, price, and production volume, all of which cannot be varied independently. Increasing the production volume may lead to adopting mass production techniques that drastically reduce the manufacturing cost and price. In most markets, a lower price can help in selling the product in larger quantities. The designer should decide on the desired manufacturing cost and production volume based on input from the finance and marketing departments. For a manufacturing firm that is starting a new product line, it is important to understand the investment needed to establish the required equipment and facilities. To this end, the firm may borrow money, and the profits generated should satisfy the payback period and return-on-investment criteria.

The targeted manufacturing cost and production volume should be established during need analysis because they will considerably affect several downstream design decisions. This kind of trade-off study was one of the critical issues in Ferdinand Porsche's historic proposal of the Volkswagen Beetle. He suggested the low target price of 900 DM ($215) for the Beetle, based on a very high production volume. Interestingly, with more than 21 million Beetles made, it remains the world's most produced car.

2.6 Size Considerations in Need Analysis

"Size" is used here as a general term to denote anything that has to do with constraints on dimensions, shape, weight, and other related physical properties. It is in this area that designers often have the most preconceived ideas, or bias, about the solution. However, any overconstraining of future solutions should obviously be avoided during need analysis.

Indeed, the natural evolution of products is such that they become lighter and more compact. In some industries, such as aerospace and electronics, product success hinges heavily on how well the designer handles the size and weight constraints. Designers are therefore under constant pressure to work within tight size constraints. These constraints in turn may eliminate certain material and geometrical choices, thereby limiting the solution space unnecessarily.

Most products must also interface with users, other equipment, and the environment. The term "environment" is used here in a broad sense to include both operating and manufacturing conditions. The interface constraints establish the required compatibility between the product and its surroundings. For instance, ergonomic constraints dictate the maximum force that can be applied by a human user. Interface constraints are often nonnegotiable because they originate from sources external to the product being designed.

2.7 Safety Considerations in Need Analysis

Safety concerns dominate almost every aspect of modern design. Accidental deaths are more common in industrialized nations than in developing or rural countries. This can be directly attributed to the greater usage of machines. With the increasing emphasis on safety, health, and the environment, new products must address not only functionality, but also safety. Safety constraints often require designing redundant and fail-safe systems.

As an example, let us look at some safety aspects in designing a car. In the case of collision, primary restraint systems (seat belts and headrests) and secondary restraint systems (airbags) are designed to hold the passengers in place. Many other components in the passenger compartment are designed so as not to inflict injury to the occupants. For example, the steering column is made collapsible. Safety aspects are also incorporated in the structural design of the car body, which needs to be strong enough to sustain rollover, and at the same time provide collapsible, energy-absorbing sections around the passenger compartment. Emission control measures and sound-attenuating exhaust systems are required to protect the health of the population at large.

2.8 Special Considerations in Need Analysis

This category is concerned with market conditions and trends, government regulations, and other, special, or unusual task-related circumstances, even including issues such as international political uncertainties. Government regulations play a crucial role in many product developments. Documents pertaining to existing regulations in the specific area of product design should be consulted during need analysis and before generating the requirements. The regulations may also specify standard tests that the product must pass before it can be released to the market. Testing standards are often more demanding than the actual service conditions and, therefore, may pose additional constraints. The following example illustrates how regulation can play a dominant role in design.

Power Source for Space Exploration

In many space missions, the technically preferred choice of power source for a spacecraft is a Radioisotope Thermoelectric Generator (RTG), a device that converts heat energy generated by decay of a radioisotope into direct current electricity. RTGs were used in several space missions including Apollo and

Galileo. However, since the 1986 Challenger accident, it has been difficult to get approval for launching a radioactive device from Earth, even though the probability of explosion in Earth's atmosphere is relatively small and the possible consequences of such an explosion are quite mild.

If the constraint "no launching of radioactive materials" is recognized, the design team will focus on developing alternative solutions that use other energy sources. If this constraint is not identified early in the need analysis, much effort can be wasted on designing a whole space mission around the wrong power source solution. This may lead to more conflicts and expensive delays. For example, NASA had to receive White House approval for the October 1997 launch (with an Earth flyby two years later) of the RTG-equipped Cassini mission to Saturn.

2.9 Development of Design Requirements

Design requirements are a list of quantitative criteria that any solution must satisfy. They summarize the need analysis and constitute the input to the next stage, conceptual design. In some industries, the requirements are referred to as "design specifications." However, it is important to note that *design* specifications or *design* requirements are different from *product* specifications. Product specifications are written at the end of the design process in order to summarize the performance and properties of the actual product. For example, the product specifications may include the actual weight, size, lowest and highest operating temperatures, and so on. In contrast, design specifications or requirements stipulate the minimum required performance and other characteristics. The requirements are written at the beginning of the design process and are used for evaluating the "goodness" of different solutions. For example, a design requirement may specify the maximum allowable weight for the design.

The requirements are a compilation of both functions and constraints in the form of short and quantitative statements. They tell *what* the system or artifact must do and *how well* it must be done. The requirements should be quantified as precisely as possible to avoid ambiguity, misunderstanding, and potential conflicts later in the design process. The magnitude of functional measures and constraints needs to be established. Obtaining the information and performing the necessary calculations are often difficult and time-consuming tasks; however, these will turn out to be useful throughout the design process. The design requirements should be consistent: there should be no conflicts between requirements, and any potential conflict must be resolved during the study of trade-offs.

The design requirements should be broad in scope to allow the consideration of innovative solutions. Sometimes, however, the set of design requirements is narrower and more focused than the need analysis. Consider the example of designing a machine for pre-recycling of aluminum beverage cans. In order to reduce the cost of storage and transportation of the cans, it is desirable to process them so that they take up less space, before shipping them to a recycling facility. The need analysis should be very broad so as not to preclude any possibility. Without actually creating a solution, a designer may look into the economical aspects of various processing technologies such as crushing, shredding/grinding, and melting. Regardless of the specific implementation, the energy required to process a single can by each technology may be estimated and compared to the cost savings in storage and transportation. Suppose it is concluded that only crushing is a viable option. The requirements developed may reflect a decision to design a crushing-type machine and may disregard the other processing techniques altogether.

The requirements need to be precise in definition and verifiable, so that they can be used later to help evaluate the design and select among alternatives. "The engine cooling system should provide sufficient cooling to prevent overheating" is a bad requirement. "Sufficient" is very vague in definition, it does not help in sizing the cooling system during design, and it does not allow quantitative comparison of alternative designs. On the other hand, the statement "The cooling system should provide a cooling capacity of 10 kW" is both precise and verifiable. Good requirements guide the design process and affect the quality of its outcome. Eventually, it is the design requirements that drive the cost and schedules of the design process, the skills and resources needed, and the verification and operational procedures used during and after design.

The designer begins developing the requirements by identifying the variables that define each function and constraint. For example, the aluminum can processing machine may require a definition of its capacity by the number of cans processed per unit time and a characterization of its output by density (i.e., the weight of aluminum per unit volume). Next, the variables are quantified by setting target values for the design, using the information obtained in the need analysis. Some variables may be assigned a single value, whereas others would have a range of acceptable values. Single values may be either maximum or minimum. To continue with the previous example, we can require that the maximum weight of the can recycling machine is 1000 N or that the minimum processing rate is 2000 cans/hour. A range of variables for this design may be assigned to the height of the loading opening, say between 1 and 1.5 meters above the floor.

Once the design requirements are formulated and quantified, they should be organized into various categories. The categories typically correspond to those of Fig. 2.5 (i.e., performance, value, size, safety, and special). Within each category, there may be one or more requirements. This organization is useful and convenient during design because it helps focus on one aspect at a time.

The design requirements list is a binding document to which every design team must conform. The requirements may be modified in the later stages of the design process; however, this would require approval from all the parties involved in the product development and may pose complications. An effort should therefore be made at the front end, that is, during need analysis, to consider all the downstream issues and compile the best possible set of requirements. On the other hand, requirements should be challenged by everyone involved in the design process, as more knowledge and understanding of the problem domain is gained. For example, consider the following scenario regarding the Mars sample return mission described earlier. The system integration group may analyze the initial need by studying old photographs of the Martian surface taken by several spacecrafts in the 1970s, and may generate a requirement for a Martian ground vehicle that is capable of traversing obstacles as big as one meter across. This requirement is dictated to the vehicle design team, which concludes that the resulting vehicle would be prohibitively large. The vehicle designers, therefore, challenge the original requirement by negotiating with the integration team, which subsequently modifies the requirement to state that the vehicle should be able to traverse obstacles that are 0.3 m in size, and handle larger obstacles by detecting and avoiding them.

2.10 Discussion and Summary

Need identification and analysis answer the following questions:

- What is the problem I'm trying to solve?
- What particular characteristics is the new product required to have?
- How will I know that I'm done?

Need identification is the step where a possibly nontechnical and sometimes configurational design task is converted to a real need. This need is extensively studied during need analysis to identify both functions and constraints in five general categories: performance, value, size, safety, and special. Throughout need analysis, the designer should be objective and careful to avoid creating solutions or developing a bias through false constraints. This will ensure maximizing the size of the solution space.

Analyzing the need also deepens the designer's understanding of the task and its domain. This will be extremely helpful later in the design process and is facilitated by being very quantitative and specific. The knowledge gained during need analysis, together with its summary as design requirements, is carried over to the conceptual design stage. Here, satisfying all the requirements serves as the criterion for completion of conceptual design. Moreover, a good set of design requirements will provide measures for the "goodness" of the design, thus facilitating conception of an *optimal* solution, not just *a* solution.

The approach presented in this chapter is a methodology for need identification and analysis. It is not a step-by-step procedure as are the following two alternative approaches. Quality Function Deployment (QFD) has been touted in recent years as a structured method of developing design requirements. It leads the designer through several steps, including listing of customers' requirements and benchmarking the competition. Alternatively, some textbooks provide comprehensive checklists of generic areas and issues that should be considered, and they ask the designer to choose the relevant points and "fill-in the blanks." Both of these approaches may be used to supplement the methodology of this chapter.

2.11 Thought Questions

1. Several design configurations may be used to bridge across a body of water, such as the English Channel or the Hudson River. Write a general need statement for this task, and examine how the need is met by bridges, causeways, tunnels, and ferries.

2. What is the real need satisfied by videocassette recorders (VCRs)? Identify competing technologies and describe the advantages and disadvantages of each one.

3. Write a need statement and list the inputs and outputs for the "black-box model" of the following products:
     (a) Clothes washer
     (b) Computer mouse
     (c) Voltage transformer
     (d) Gearbox
     (e) Automobile brakes
     (f) Automobile tires

4. The evolution of tires involved several significant changes:
     (a) From bare metal wheels to solid rubber tires
     (b) From solid rubber to pneumatic tires with inner tubes

    (c) From tubed tires to tubeless bias-ply tires

    (d) From bias-ply tires to bias-ply belted and then to radial-ply tires

Investigate the evolution of tires to determine the needs satisfied by the different configurations and the performance trade-off issues in their design. What are the performance requirements for an ideal automobile tire?

5. Identify the trends in the laptop computer market with the goal of developing a new laptop computer in the next year. Using these trends, determine a target weight, size, screen size, and resolution for the new laptop.

6. Identify the important safety considerations in the design of a food processor.

2.12 Bibliography

Quality Function Deployment is widely described in the literature. A popular paper is:

> Hauser, J. R. and Clausing, D. "The House of Quality." *Harvard Business Review,* 66, No. 3 (1988): 63–73.

A more extensive coverage of QFD can be found in:

> Clausing, D. *Total Quality Development.* New York: ASME Press, 1994.

The following two textbooks use QFD as a tool for understanding the design task and developing the requirements (dubbed "engineering specifications"):

> Dixon, J. R. and Poli, C. *Engineering Design and Design for Manufacturing: A Structured Approach.* Conway, MA: Field Stone Publishers, 1995.
>
> Ullman, D. G. *The Mechanical Design Process.* 2nd ed. New York: McGraw-Hill, 1997.

The use of checklists is demonstrated in:

Pahl, G. and Beitz, W. *Engineering Design: A Systematic Approach.* London: The Design Council, 1988.

3

Need Identification and Analysis Case Study: Packing Factor of Sand in Electrical Fuses

In sand-filled electrical fuses, the packing factor of sand controls the overall fuse performance. This chapter discusses an ill-defined, *perceived* need involving the packing factor. The methodology outlined in the previous chapter is applied to define the *real* need by establishing the scope of the design task at the proper level. The need is then analyzed in order to fully understand and quantify the key issues. The results of the need analysis are summarized as a set of design requirements. The case study demonstrates a definitive link between good need identification and analysis and the potential for innovation.

3.1 Background

Sand-filled fuses are used to protect electric mains and feeders, circuit breakers, heating and lighting circuits, motors, transformers, semiconductors, and more, against current surges. They are characterized by their interrupt rating, which is the maximum current that a fuse can stop while maintaining its mechanical integrity. Fuses that have high interrupt rating are filled with sand. A schematic of a sand-filled fuse is shown in Fig. 3.1. The components are:

- *Fuse element:* made of silver with one or more weak spots depending on the interrupt rating.
- *End-caps:* connect the fuse in an electrical circuit. A hole is provided in one of the end-caps to allow sand filling. The fuse

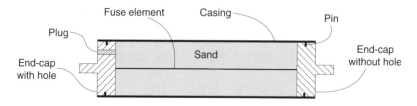

Figure 3.1 Schematic of a sand-filled fuse.

element is soldered to the end-caps. The external protrusions are the electrical terminals of the fuse.

- *Sand:* fills the space in the casing around the fuse element.
- *Fiberglass casing:* houses the fuse element and the sand.
- *Plug:* closes the hole in the end-cap after sand filling.
- *Pins:* attach the end-caps to the casing.

When the current in the circuit exceeds the prescribed limit, the weak spots in the fuse element are designed to melt and create a discontinuity. An arc forms across the discontinuity and continues to conduct current. During this arcing process, the sand serves two key purposes:

1. It melts due to the intense heat of the arc and absorbs energy equivalent to the heat of fusion. As a result, it diminishes the amount of energy available for heating the gases in the casing and thereby reduces the possibility that the casing will explode due to expanding gases.
2. The porosity of the sand filling allows the molten metal to escape from the arcing interface. This reduces the time to interrupt the circuit.

Thus, the sand filling helps to maintain the mechanical integrity of the fuse while reducing the interrupt time.

3.2 The Initial Need

The time to interrupt the circuit is a key performance feature of the fuse. A short interrupt time is desirable because it decreases the load on various components during the short-circuit condition. The time to interrupt the circuit is highly dependent on the packing factor *(PF)* of sand. The *PF* is defined at the macroscopic level as the fraction of space occupied by sand grains in the fuse cavity. The general expression for the *PF* is:

$$\text{Packing Factor} = \frac{\text{Volume of sand}}{\text{Volume of fuse cavity}}$$

A typical sand-filled fuse has a *PF* of 0.63 to 0.68. As the *PF* decreases, the interrupt time increases and becomes unacceptable at 0.61 for most fuses. A fuse manufacturer asked that a quality control device be designed and incorporated in the production line to identify faulty fuses. The stated initial need was to "check the *PF* of each fuse before shipment."

3.3 Need Identification

The black-box approach can be used to study and define the real need. For this particular task, the desired output is the *PF*. The fuse is both the input and the output for the black box as shown in Fig. 3.2. Based on this model, the need can be formulated as "determine the *PF* of the fuse."

Now we must ask ourselves whether the need statement is not too constraining; that is, whether it will permit us to conceive innovative solutions. This particular need statement suggests that the *PF* be determined *after* the fuse is completely assembled. This limits the range of possible solutions to just a few nondestructive evaluation techniques. On the other hand, if the implied constraint "the *PF* must be determined only after assembly" is removed, then the designer is free to conceive more innovative solutions. For instance, the *PF* can be determined based on measurements taken *during* the manufacturing process. Thus, the same need can be formulated at a more generic level as "determine the *PF* during the manufacturing process."

If we can determine the *PF* before closing the end-cap with the plug, then it may be possible to add more sand to achieve the desired *PF*. This raises the interesting question of whether we want merely to *find* the *PF* or whether we actually want to *control* it. The latter

43

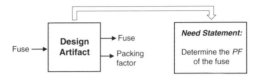

Figure 3.2 Initial black-box model of the design task and the resulting need statement.

choice expands the scope of the design task, so the new need can be stated as "ensure the correct *PF* during the manufacturing process." A black box for this particular need is shown in Fig. 3.3.

3.4 Need Analysis

We first list relevant questions that come to mind, such as:

- Can the *PF* be determined from the weight or mass of the fuse?
- What are the sources of variation of the *PF*?
- What is the ideal range for the *PF*?
- What are the production rate and volume of the fuses?
- What is the cost of a fuse?
- How much would the fuse manufacturer be willing to pay for the designed equipment?
- How large can the equipment be?

Next, we attempt to develop a thorough understanding of the functions and constraints involved with this design task, under the five categories of *performance, value, size, safety,* and *special.*

Performance

The primary function of the design is to ensure the correct *PF*. For fulfilling this goal, it is important to identify:

- The maximum possible value for the *PF*.
- The design variables that determine the *PF*.
- The primary sources of variation in the magnitude of the *PF*.

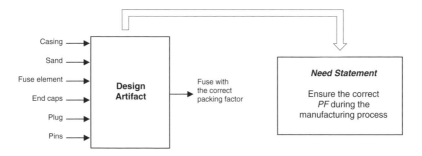

Figure 3.3 Revised need definition for the design task.

To determine the *PF,* we can assume the sand grains to be incompressible spheres. From material science we know that the *PF* is dependent on the spatial arrangement and independent of the grain size; that is, two different grain sizes with the same spatial arrangement will give the same packing factor. For any grain size, the *PF* can range from 0.52 for a simple cubic structure to 0.74 for an hexagonal close-packed structure. However, including interstitial grains can increase the *PF.*

Now let us understand the design variables that determine the *PF.* As discussed earlier, the *PF* is the ratio of the volume of the sand to the volume of the cavity. The volume of the sand is proportional to its mass. The volume of the cavity depends on the inside diameter of the casing and the length of the cavity. By substituting terms in the general expression for *PF* we get:

$$PF = \frac{\text{(Mass of sand/Density of sand)}}{\text{Volume of fuse cavity}} = \frac{(M/\rho)}{(\pi d^2/4) \times l}$$

where M is the mass of the sand, ρ is its density, d is the inside diameter of the casing, and l is the length of the cavity.

The mass of the sand can be determined by subtracting the mass of all the components from the total mass of the fuse. However, due to manufacturing tolerances, the mass of the fuse components can vary significantly, as shown in Table 3.1. Similarly, dimensional tolerances on the casing and the assembly process can cause the cavity volume to change. These variations are summarized in Table 3.2.

45

Table 3.1 Mass variation of the components of sand-filled fuses.

Component	Mean (g)	Standard Deviation (g)
Fuse element	0.9	0.0025
End-cap without hole	37.75	0.175
End-cap with hole	36.25	0.175
Fiberglass casing	19.00	0.05
Plug	1.25	0.002
Pins	1.00	0.005

Table 3.2 Cavity diameter and length variation.

Design Variable	Mean (mm)	Standard Deviation (mm)
Diameter	8.00	0.01
Length	26.3	0.1

The two tables reveal that if the total weight of the fuse is used for determining the *PF*, then the variation in the mass of the end-caps and in the length of the cavity can result in significant error. The mass variation of other components and the casing diameter variation result in much smaller errors.

Value

The value of the new system would be the increased quality of fuse production. Quality Loss Function (QLF) is a systematic and rational methodology for estimating loss of quality due to off-target performance in monetary terms. This methodology will be used here to analyze and quantify the value of the new design.

We know that ensuring a good *PF* results in high-quality fuses. As the *PF* deviates from the desired value, the quality of the product decreases, and this, in general, manifests itself as a loss to society. These losses can be due to the product failing to deliver on-target performance, to harmful side effects of the product, and to down-time of the equipment. Quality loss is often represented by the following quadratic function:

$$L(y) = ky^2$$

where $L(y)$ is the loss function, k is a constant known as quality loss coefficient, and y is the difference between the desired and actual PFs. For this problem, we can assume the desired PF to be 0.74. The quality loss of a fuse is zero (i.e., no loss in quality) when the actual PF is equal to the desired value. As the PF deviates from the desired value, the quality loss increases. Customers take economic action when the quality loss becomes unacceptable to them. The customer tolerance Δ_0 is defined as the value of y at which half of the customers would take economic action. Based on an informal survey of fuse customers, half of the customers who have fuses with a PF of 0.61 would switch to another manufacturer. Therefore, the customer tolerance was set at 0.13 (desired PF of 0.74 minus actual PF of 0.61).

To quantify the loss, we need to estimate the value of the quality loss coefficient k. This can be done by identifying the loss A_0 at the customer tolerance of the PF. In reality, the estimation of loss is quite complex since the cost should include the downtime and the replacement of damaged equipment for a typical application. For simplicity, it is assumed that the estimated average replacement cost (including the cost of a new fuse and the installation cost) is $2. Thus, the QLF for this problem is:

$$L(y = \Delta_0) = k\Delta_0^2 = A_0$$

$$k = \frac{A_0}{\Delta_0^2} = 118.3$$

and the quality loss of a fuse with packing factor PF can be written as:

$$L(y) = 118.3 \times (\text{Desired } PF - \text{Actual } PF)^2$$

Our interest is in finding the average loss for a given set of fuses manufactured by the current production process. The average quality loss can be computed as:

$$\bar{L}(y) = \frac{\Sigma(\text{Quality loss of each fuse})}{\text{Total number of fuses}}$$

$$\bar{L}(y) = k \frac{(y_1^2 + y_2^2 + y_3^2 + \cdots + y_n^2)}{n}$$

$$\bar{L}(y) = \frac{k}{n} \sum_{i=1}^{n} y_i^2 \approx k(S^2 + \bar{y}^2)$$

where $\bar{L}(y)$ is the average quality loss, n is the number of fuses, $\bar{y}$ is the mean y (the mean difference between the actual and desired PFs), and S^2 is the variance of the PF. The last equation shows that the quality loss can be decomposed into two parts:

1. Quality loss due to the mean being off target.
2. Quality loss due to the variance.

The quality loss can be reduced by bringing the mean PF close to the target (0.74) and reducing the variance. Figure 3.4 shows the same insight: the average quality loss decreases by increasing the PF or by reducing the variance.

By substituting values for the fuses manufactured through the current system ($PF = 0.66$, $\bar{y} = 0.08$, and $S^2 = 0.01$), the average quality loss can be calculated as $1.94. The target is set at $PF = 0.67$, $\bar{y} = 0.07$, and $S^2 = 0.005$. The average quality loss at this target is $1.17 from Fig. 3.4. Thus, the target performance will result in an increased value of $0.77/fuse ($1.94 − $1.17). Based on a production volume of 100,000 fuses per year, the new system would reduce the losses due to bad quality by $77,000 in the first year. Even though this is not a cash flow to the manufacturer, it is realized in terms of greater customer retention and increased customer satisfaction. A target cost for the equipment was set at $40,000 based on the possible quality improvements by the new system. Note that the relatively low production volume relates to the particular size of the high-performance fuse.

Size

Because of the tight space constraints in the plant, the maximum floor space available for the equipment is 8' × 5'. The maximum permissible height is 8'.

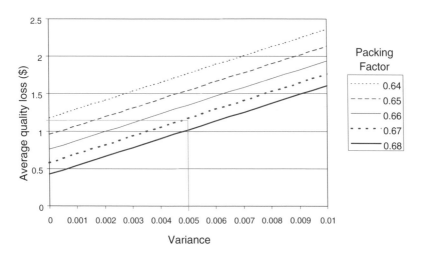

Figure 3.4 The average quality loss for different combinations of *PF* and variance.

Safety

Any device on the shop floor should meet the relevant Occupational Safety & Health Administration (OSHA) regulations. These regulations often make some solutions less practical. For instance, this requirement may rule out the possibility of using an X-ray machine on the shop floor to determine the *PF*.

Special

Any modification to the fuse design should meet Underwriters Laboratories (UL) regulations. The UL regulations set limits on several factors such as I^2t (proportional to the energy let through the system).

3.5 Design Requirements

Table 3.3 summarizes the requirements generated in each of the need analysis categories.

Table 3.3 The design requirements list.

Category	Design Requirement
Performance	1. The device must ensure a minimal packing factor of 0.62.
	2. Target packing factor: 0.67 mean with 0.005 variance.
	3. Mass specification of components:

Component	Mean (g)	Standard Deviation (g)
Fuse element	0.9	0.0025
End-cap without hole	37.75	0.175
End-cap with hole	36.25	0.175
Fiberglass casing	19.00	0.05
Plug	1.25	0.002
Pins	1.00	0.005

4. Current cavity specification:

Design variable	Mean (mm)	Standard Deviation (mm)
Diameter	8.0	0.01
Length	26.3	0.1

5. Sand: 460 grade
6. Production volume: 100,000/year
7. Production time: 1.5 minute/fuse

Value   8. The equipment cost should be no more than $40,000.
Size   9. Available floor space: 8′ × 5′
10. Maximum height: 8′
Safety   11. Must meet OSHA regulations for safety.
Special   12. Any modification to the fuse design should meet UL regulations.

3.6 Discussion and Summary

During the need identification step, the need statement changed from "determine the *PF*" to "ensure the correct *PF*." Such expansion of the scope increases the solution space and the probability of finding an innovative solution, as discussed in this section.

The first need statement, "determine the *PF* of the fuse," may result in solutions that nondestructively evaluate the *PF*. One such solution uses acoustic emissions. When a fuse is shaken, the sand grains impact each other and the housing. In the process, they produce sound. The intensity of acoustic emission is inversely proportional to the *PF*. In other words, a loosely filled fuse emits more noise. This energy measurement can be correlated with the *PF*.

The modified need statement, "ensure the correct packing factor during the manufacturing process," expands the solution space to include various measurements *during* manufacturing, such as the length of the cavity during assembly and the mass of the sand during the filling process. Based on the insight that there is very little variation in the cavity diameter, these measurements can be used to determine the *PF*.

Even more innovative designs are facilitated by shifting the emphasis from quality control to process modification. Identifying the sources of variation in packing factor may lead to a simple, yet innovative, solution involving a fixture to control the length of the cavity. Since the volume now becomes almost constant, the focus of the design task is shifted to filling a predetermined quantity of sand.

Since we learned that changing the uniform grain size has very little effect on the *PF*, we can focus on introducing a second grain size to fill the interstitial voids. This can increase the *PF* without significantly affecting the manufacturing process. In summary, a good need identification coupled with a good need analysis results in crucial insights that can translate into innovative design solutions.

3.7 Thought Questions

1. Quantify the sensitivity of the packing factor to various design variables.

2. Determine the accuracy of computing the *PF* by weighing the fuse.

3. Identify three different ideas that may be used in a design to "ensure the *PF*."

3.8 Bibliography

The following book is a general reference for fuse design:

Wright, A. and Newbery, P. G. *Electric Fuses.* 2nd ed. London: Institution of Electrical Engineers, 1994.

Quantifying the loss of quality due to off-target performance is discussed in:

Fowlkes, W. Y. and Creveling, C. M. *Engineering Methods for Robust Product Design: Using Taguchi Methods in Technology and Product Development.* New York: Addison Wesley, 1995.

4

Introduction to
Parameter Analysis

This chapter presents the systematic methodology for conceptual design, namely, parameter analysis. First, we look at an invention and try to reconstruct the designer's thoughts that led to it. We explore the nature of the conceptual design process and explain it as continuous movement between concept space and configuration space. This theoretical model is used to demonstrate the invention described earlier and to outline the parameter analysis methodology for conceptual design.

4.1 A Look at an Invention

A tiltmeter, or inclinometer, is a device capable of measuring extremely small angles of tilt with respect to the local gravity vector. When used for sensing "earth tide," the distortion of the Earth's crust due to the gravitational pull of the moon, the tiltmeter needs to measure changes in angle of the order of 10^{-6} radians. Another use is in predicting earthquakes, where the required resolution is of the order of 10^{-7} to 10^{-9} radians. Figure 4.1 is a schematic representation of a tiltmeter that was invented in the 1970s. The lines represent stiff members, the circles are the pendulum weights, and the solid dots are hinges. A small tilt, α, given to the base disturbs the equilibrium of Fig. 4.1a and produces a much larger, and therefore easier to detect, inclination β of the pendulums, Fig. 4.1b.

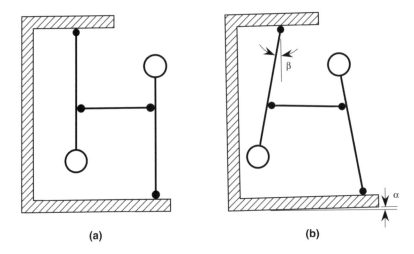

Figure 4.1 (a) Schematic of the tiltmeter with no input angle and (b) the tiltmeter's large response angle β when measuring a small tilt angle α (not to scale).

It is fairly obvious, as it was to the inventor, that a simple pendulum (Fig. 4.2) is a configuration that provides a measurement of tilt angle through a measurement of the lateral displacement of the pendulum weight. However, the size of the device presents a problem. The lateral displacement of the weight at the bottom of a simple pendulum is large enough to be measured at small angles of tilt only when the pendulum is extremely long. For example, a 50-m long pendulum would produce a displacement of 5 μm when the angle is 10^{-7} radians. This is a relatively obvious concept. However, the inventor also realized that one could represent this displacement relationship as a simple spring. A displaced spring pushes back with a force f, which is proportional to the displacement Δx ($f = k\,\Delta x$). For the pendulum of Fig. 4.2, the restoring force f is also proportional to the lateral displacement:

$$f = mg \sin \theta = mg\,\frac{\Delta x}{l} = \frac{mg}{l}\,\Delta x = k\,\Delta x$$

Therefore, the statement that the pendulum must be very long is the same as saying that k, the stiffness of the pendulum, must be

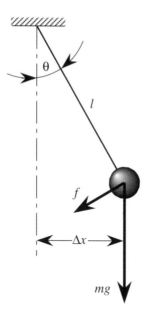

Figure 4.2 A simple pendulum with a restoring force *f* may be viewed as a spring.

very small. Continuing this logic, the inventor then recognized that there is another way to obtain a small spring constant in addition to extending the length of a simple pendulum. The difference between two large spring constants (short pendulums) can yield a small spring constant (effectively long pendulum). This relationship is represented by $f = (k_1 - k_2)\,\Delta x$, but it requires a negative spring in order to obtain the $-k_2$ term. A negative spring would "pull" in the direction of the initial displacement with a force proportional to the displacement. Many statically unstable devices exhibit a similar behavior; for example, some light switches "jump" to a different state when pushed slightly. The inventor of the tiltmeter noted that an inverted pendulum is a negative spring. Thus, all that remained was the coupling between the two pendulums at a point at which the resultant spring constant $(k_1 - k_2)$ is sufficiently small but positive to yield very high sensitivity (large lateral displacement) for small angles of tilt.

This very brief description represents only a small portion of the process that took place in creating the new tiltmeter. However, it does describe the kernel events around which the whole concept revolves. We shall return to the tiltmeter later, but let us first examine a few fundamental issues relating to conceptual design and attempt to establish some principles with regard to the process.

4.2 The Nature of Conceptual Design

When we look at inventions such as the tiltmeter, we often ask basic questions such as:

- Why is conceptual design so difficult?
- Where do new ideas come from?
- Why didn't I think of that?

These questions seem almost trivial or trite, but they are fundamental to conceptual design, and an attempt to answer them helps to elucidate the subject.

Let us first consider why conceptual design or the generation of new ideas is such a difficult task. Implicit in this question, of course, is the task of generating *very good* ideas as opposed to just generating ideas. It is quite a simple matter to generate many, many ideas when the demands on you do not include the quality or "goodness" of the ideas. As we all know, many more ideas are generated than are developed; many more are developed to some level of completion than are made into products; and many more products are made than are successful. The real issue comes down to generating ideas of high quality and of good value to others.

Figure 4.3 is a schematic view of why conceptual design is fundamentally more difficult than many other stages in the design process. The result of conceptual design is, hopefully, a new concept or configuration. It is this "newness" that makes the process so elusive. A new configuration that heretofore did not exist has been created, and the process of reaching that point is very different from the activity that takes place during the refinement of the concept downstream in the process. The horizontal axis in Fig. 4.3 is a one-dimensional representation of the many

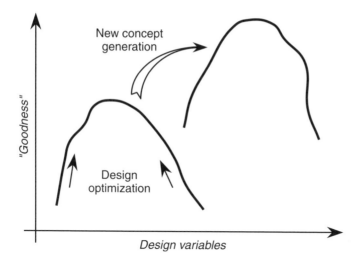

Figure 4.3 The relationship between conceptual design and design optimization.

design variables that can be changed within a configuration. The vertical axis is a one-dimensional representation of the quality or "goodness" of a concept. Such a measure may be imagined to be the ratio of performance to cost since, fundamentally, we would all like to pay less for more performance. In any event, such a measure is a convenient way to describe the desirability of concepts or configurations.

Consider now the curve on the left and imagine that this curve is actually a multidimensional surface in space and that the points on this surface represent different combinations of the design variables that can be reached. All points on the surface are various realizations of a single concept.

At this point, two simple examples will help. Imagine that the curve on the left describes a fountain pen. That is, the core technical concept is described by the fact that the pen contains a nib through which ink flows by capillary action from some sort of a reservoir to the paper. Points on this surface may describe pens that range from a quill pen to a sophisticated refillable fountain pen to a fountain pen that has a very wide tip used for calligraphy. These pens differ in the values of many design variables, but the core technical concept is still the same. Movements from one point to others on the same surface may come relatively easily by making minor changes to a design variable. These changes might include modifying the type or size of the reservoir of ink, changing the physical dimensions of the nib, or making small changes to the properties of the ink. All of the realizations are still on the surface because they all share the same core technical concept. Each of the points represents one realization of the common attribute of "being a fountain pen." These movements may be described as *design optimization*—a selection of certain existing design variables in order to meet a performance specification.

In contrast, *conceptual design* is represented by a leap from one surface in the design space, the left curve in Fig. 4.3, to a new surface, represented by the curve on the right. The new curve may not even be described by all of the same design variables. Before the leap was made, the new core technical concept did not exist. To extend the example, imagine that the new curve represents a ballpoint pen. In fact, some of the design variables that are pertinent to the fountain pen are also important to ballpoint pens. However, these two configurations do not share a number of design variables, and there is a fundamental difference between the core technical concepts of a fountain pen and a ballpoint pen. The fountain pen uses capillary action to feed the ink while the ballpoint pen does it through the mechanical rolling motion of the ball. All points on the curve to the left are fountain pens, all points on the curve to the right are ballpoint pens, and the generation of the ballpoint pen concept was not obvious in the consideration of the fountain pen designs. It is the fact that this process is a leap to a nonexistent curve that makes conceptual design so difficult.

As a second example, suppose the left curve in Fig. 4.3 describes the concept of reciprocating internal combustion engines, while the right curve represents rotary engines. The reciprocating engine curve would consist of points representing various configurations (V-shaped, in-line, horizontally opposed, radial), number of cylinders, methods of valve actuation (side valves, overhead cams), and so on. However, all these configurations would be based on the concept of reciprocating pistons in cylinders, connecting rods, and crankshafts (Fig. 4.4a). The rotary engine curve may represent different geometries (for example, a triangular rotor in a figure-eight housing as in the Wankel engine, Fig. 4.4b, or any of the many other theoretical possibilities), numbers of rotors, and so on. Again, movement along the same curve stands for changing some of the design variables, such as the number of spark plugs used or the shape of the combustion chamber. In contrast, the creation in the 1950s of the concept of rotary engines is denoted by the establishment of the new surface on the right in Fig. 4.3 and represents a conceptual breakthrough.

This discussion of the generation of new ideas has been only philosophical, but it should help to clarify why the creation of truly new, high-quality configurations or ideas is so difficult and therefore an event too rare in engineering practice today. It also shows that good ideas do not usually emerge from doing conceptual design in the form of a search over existing solutions. Rather, conceptual design is a discovery process.

Let us now try to answer the other two questions posed at the beginning of this section. Study of the process of conceptual design and a look at many inventions allow us to make the following observations:

- Human creativity is more successful or productive when one is attempting to solve simple problems rather than complex ones.
- The best ideas are usually quite simple conceptually.

These notions need some explanation. The first notion deals primarily with the number of issues or aspects that must be considered during the solution process. Thus, low-order problems, those that require consideration of very few parameters or factors, tend to be much more easily solved than higher order problems, those with a large number of variables to be considered. That is why all of us try to break down problems into smaller pieces in an effort to make them more tractable. This simple notion comes as no surprise but has profound implications for the nature of successful creative processes.

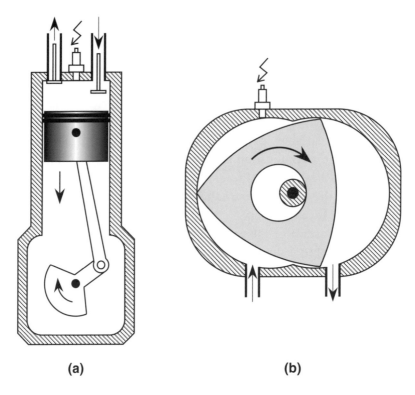

(a)   (b)

Figure 4.4 (a) A reciprocating internal combustion engine and (b) a rotary Wankel engine.

The question, "Where do good ideas come from?" may be addressed by examining the second notion, which relates to the simplicity of the results of the creative process. It is not that the resulting products based on these good ideas (core technical concepts) are simple, rather, the concepts themselves are most often based on simple but creatively new insights. Often, these simple ideas can be described as resulting from transformations of the key issues in a problem into new relationships or sets of relationships uniquely identified by the creative designer.

As a result, the answer to the frequently asked question, "Why didn't I think of that?" may be twofold: Perhaps I failed to break down the task in a way that would allow me to focus on one issue at a time; or, while in retrospect the ideas behind the design are simple, the pathway to reach them may not have been at all obvious to me.

4.3 Theoretical Model of Conceptual Design

The notions regarding the nature of conceptual design lead to several conclusions about a model for the conceptual design process. Figure 4.5 presents a simple schematic model. The output of the process is generally a physical object or group of objects that here are called *configurations*. Thus, the various outputs as the process unfolds can be thought of as elements in *configuration space*. However, consideration of the preceding observations concerning problems and solutions within conceptual design indicates that movement from one point in configuration space to another point in configuration space is generally made at the level of *concepts*. Therefore, the model also contains another space, called *concept space,* which contains the ideas or concepts that provide the basis for the elements of configuration space. The conceptual design process can be viewed as an iterative process of moving from configuration space to concept space, making changes or movements within concept space, and then moving back to new points within configuration space corresponding to further creative generation of configurations based on these new ideas.

The process of moving from configuration space to concept space can be thought of as *abstraction* or *generalization*—that is, generalizing from the particulars of a physical configuration to a conceptual understanding in order to change it or improve upon it. Movement from concept space to configuration space can be regarded as *realization* or *particularization*—bringing to reality, in particular physical form, the technical concepts arrived at within concept space. Readers should note that in this process description, the elements of configuration space are not actually physical objects but only diagrams, sketches, or other representations of physical objects motivated by the conceptual elements of concept space, which bring some real form to the thoughts created in concept space.

This model should be viewed as much more than a philosophical statement. Readers are encouraged to examine their own experiences in conceptual design to be further convinced of its significance. As ideas come to life, there are many points along the way which are either intermediate configurations (elements in configuration space) or conceptual statements or relationships (elements in concept space). Rarely

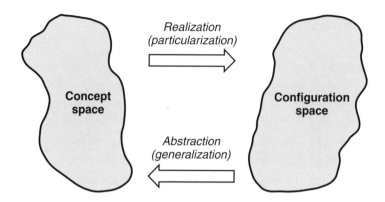

Figure 4.5 A theoretical model of conceptual design.

within this process is a new configuration created from the previous one without an excursion to concept space, with new conceptual insight being the driving force for the new configurational result.

Let us return now to the tiltmeter example. Of particular importance is the nature of the events that took place within concept space. First, there was *simplification* of the problem by considering only the pendulum configuration and ignoring, temporarily, many other issues that obviously are important. Second, there was a *transformation* from the normal way of looking at a pendulum to viewing a pendulum as a spring. Third, there was the creative step of recognizing the relevance of the difference between two large numbers to the situation at hand. Finally, we should not ignore the additional creative steps to generate the double pendulum configuration.

Figure 4.6 is a modification of Fig. 4.5 in order to include an important observation. On the right-hand side, within configuration space, are represented two realizations of pendulum devices: a simple pendulum and the new coupled-pendulum tiltmeter. On the left-hand side, within concept space, are two concepts: the simple spring relationship and the spring relationship represented by the difference between two stiff springs. The transition from the simple pendulum to the complex pendulum within configuration space is not likely to take place by itself. On the other hand, the transition from the concept of a simple spring, correlating with the properties of a pendulum, to the second concept is motivated by recognizing

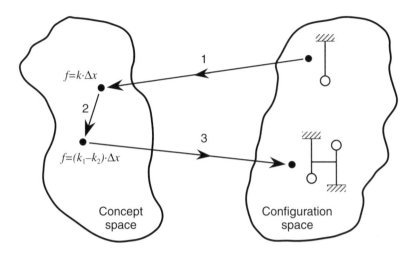

Figure 4.6 A simplified map of the creation of the new tiltmeter.

the elements of these conceptual issues. Thus, the creative motion within configuration space is actually a movement driven by motion within concept space. In simple terms, the process unfolds with the movements as labeled numerically in the figure.

Notice also that the movement labeled 1 was in itself a creative step. It represents a simplification and transformation of the typical understanding of pendulums to a form that, downstream, led to a very interesting result.

The reader should be aware that this diagram of movements is extremely simplified. In this example of creative conceptual design, many more motions were taking place back and forth between concept and configuration which have not been included in the diagram. Furthermore, the process at this stage of the description is far from over. Many more conceptual issues remain, such as how to obtain hinges with the properties necessary to allow the new device to perform as required and how to measure the lateral displacement of the pendulum weight.

Of course, conceptual design problems have many solutions, as does this problem of measuring extremely small angles of tilt. For example, Fig. 4.7 shows three other configurations that could be used instead of the double pendulum to implement the concept of measuring tilt with a long pendulum. The virtual length of the

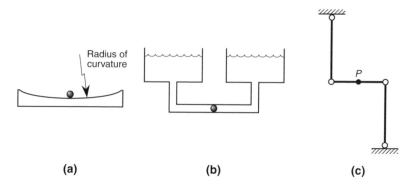

Figure 4.7 Three realizations of the "long-pendulum" concept: (a) a ball rolling on a slightly concave surface, (b) a ball moving in a pipe connecting two reservoirs, and (c) a mechanism whose point *P* moves along a trajectory with infinite radius of curvature.

"pendulum" of Fig. 4.7a equals the radius of curvature of the surface on which the ball rolls, which can be made very large. The configuration of Fig. 4.7b uses a large difference in area between the two reservoirs and the connecting pipe to obtain a large gain. The movement of the small ball, or the interface between two nonmixing fluids, is used to detect the tilt. Figure 4.7c is only one of many examples of mechanisms that have a point whose trajectory has an infinite radius of curvature. The significance of the double pendulum example is not the particular solution presented but the types of information that were used during the process and the relationship between the elements as the process moved forward.

Readers will remember instances when they were "stuck in a rut." One may represent this situation as an effort to undergo creative movements within configuration space, when in fact it is most beneficial to make significant movement within concept space and between concept space and configuration space.

4.4 Parameter Analysis as a Conceptual Design Methodology

Understanding the nature of the conceptual design process and the process model proposed above leads us to generate a methodological approach to carrying out conceptual design. We must first state very clearly that this approach is not to be considered a method but rather a methodology. The importance of this understanding cannot be minimized since the word "method" implies that a procedure that unfolds in an orderly and sometimes predictable manner results from such an approach. However, successful design methodologies must be much more open, iterative, and flexible than a highly systematized method would allow. Therefore, the methodology presented here is a general approach based on the fundamental notions regarding the nature of conceptual design. A methodology gives one the general direction of how to approach problems rather than outlining a rigid procedure.

Readers should also appreciate that the structured framework in which the methodology is presented serves only to describe the approach in an effective way. Thus, although the methodology appears in a very graphical and orderly form, this is only to illustrate the general principles on which the methodology is based.

A general word about problem solving techniques is in order. As a way of describing various techniques to enhance creativity and, in particular, conceptual engineering design, one might describe various approaches as being "environmental" or "methodological." What is meant by environmental here is that creativity is encouraged or stimulated by establishing an environment in which creativity is more likely to occur. Methodological approaches, such as that which is described here, consider the enhancement of the creative process from the point of view of active intervention in the unfolding of the process rather than of simply causing conditions to be met that will encourage creativity.

The methodology for conceptual design presented in this book is called parameter analysis and is represented by the schematic of Fig. 4.8. Figure 4.9 indicates the correspondence between the theoretical description of the process and this methodological approach. The three types of activity within parameter analysis form an iterative loop. The first element, *parameter identification,* consists primarily of the recognition of the dominant parameters or issues in a problem. The word "parameter" is used to describe in a very general way any issue, factor, concept, or influence that plays an important part in developing an understanding of the problem and pointing to potential solutions.

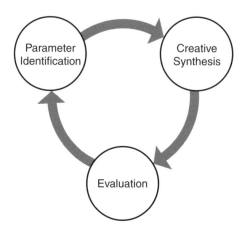

Figure 4.8 Schematic of the parameter analysis methodology.

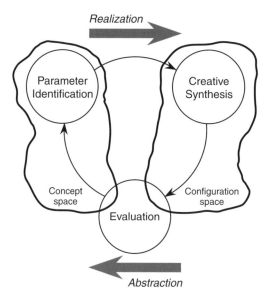

Figure 4.9 Correspondence between the theoretical model of conceptual design and the parameter analysis methodology.

The best descriptions of what takes place within parameter identi-fication are simplification and transformation. The parameters within a problem are not fixed but rather are those that are creatively recog-nized as the process moves forward. The recognition of interesting new parameters, based on transformations of more commonly understood parameters, triggers new, unique solutions to problems. Also, temporary purposeful oversimplifications of a problem, made to focus on only one of the key parameters that characterize it, help to concentrate creative activity so that it can produce quality ideas. Identification of these key parameters through simplification and transformation does not yield results unique to a problem but is a highly creative process.

The second part of parameter analysis is *creative synthesis.* This part of the process represents the generation of a physical configuration based on a concept recognized within the parameter identification step. Since the process is iterative, one should expect many physical configurations to be generated as the process unfolds, not all of which will be very interesting. However, the physical configurations allow one to see new key parameters, which will yet again stimulate a new direction for the process.

This process of moving from a physical realization back to parameters or concepts is facilitated by the third part of the parameter analysis methodology, the *evaluation* step. Evaluation is important because one must consider to what degree a physical realization is a possible solution to the entire problem. As parameter identification and creative synthesis proceed, one should expect many physical configurations to be generated. While these are far from being valid solutions to the problem, nevertheless, they are important to the stimulation of the creative process. Evaluation should not contain analysis of physical configurations that is any deeper than that required to create a fundamental understanding of its underlying elements. We might best call this process "appropriate" analysis. It must be carried out in the light of the entire need, not just with respect to the particular simplified parameter of each iteration.

The major implication of the theoretical model that stands behind parameter analysis is that the burden of truly creative activity is shifted from what here is called *creative synthesis*—the generation of physically realizable configurations—to *parameter identification,* the creation of new conceptual relationships or simplified problem statements (all lumped under the term "parameters"), which will lead to the desired configurational results. Thus, the task of creative synthesis along the way is only to generate configurations that, through evaluation, will enlighten the creative identification of the next interesting conceptual approach. Each new configuration does not have to be a good solution, only one that will further direct the discovery process.

4.5 Discussion and Summary

Several characteristics are at the foundation of parameter analysis:

- Behind every configurational attribute of a design there must be a conceptual reason or "justification."
- Developing a design requires focusing on one or very few critical conceptual issues at a time.
- The most creative aspect of conceptual design is not the invention of a new configuration, but rather the designer's ability to discover a conceptually new way of looking at the problem.
- The key to good conceptual design is a thorough understanding of the underlying physics of design artifacts.
- Conceptual design is not pure synthesis; rather, it consists of a partnership between synthesis and analysis.
- Good design is a synthesis of a series of good ideas, or concepts, not just one good idea. The parameter analysis methodology helps to create ideas and incorporate them in a single design.

Conceptual design by parameter analysis can be done at different levels of abstraction when considering the degree of innovation involved. The distinction made in Section 4.2 between design optimization and new concept generation depends on the scope of the particular design. For example, the invention of the rotary Wankel engine may be perceived as the creation of a new concept (and the generation of a new surface in Fig. 4.3) in the broad context of designing internal combustion engines. At this level, the invention of the spherical piston head, as opposed to flat-top pistons, is merely an optimization step illustrated by movement along an existing surface to a point with a higher "goodness" value. However, had the scope of the design task been limited to improving the efficiency of reciprocating engines, the degree of innovation introduced by spherical piston heads would certainly entail the creation of a new surface or concept. Similarly, the leap from external to internal combustion engines (and the future leap from internal combustion engines to batteries and fuel cells) is a bigger innovation than the move from reciprocating to rotary engines.

Parameter analysis can also be applied to "routine" design tasks, those that we would not normally associate with innovation. Many designs are developed through a process that is more evolutionary than revolutionary. Most developments in the automotive industry, for example, are of this nature. This does not imply that today's cars are not good designs. On the contrary, and perhaps thanks to the long evolutionary development of automobiles, they represent a very solid and refined solution to the basic need of providing a means of transportation. Evolutionary designs do not necessarily lack innovation. Electronic ignition systems, fuel injection, antilock brakes, and airbags are all novel designs. And parameter analysis can be applied to refining an existing design as much as to tasks that are totally new and do not have any existing solution.

4.6 Thought Questions

1. Find additional examples of *design optimization* versus *new concept generation*, as described in Section 4.2. How do these examples depend on the *level of innovation*, as discussed in Section 4.5?

2. Study the operation of felt-tip pens. Do they represent a design optimization on an existing fountain pen or ballpoint pen "curve," or do they constitute a new concept?

3. Look back at a design process, either yours or someone else's, and classify its elements (e.g., decisions, ideas, sketches) into those that belong in concept space and those that are members of configuration space.

4. The tiltmeter described in this chapter requires near frictionless hinges. Find out how this is done in other applications with sim-

ilar demands. Then, check what is suggested for this purpose in the original tiltmeter patent (see Bibliography).

4.7 Bibliography

Much of the material in this chapter is also found in:

Jansson, D. G. "Conceptual Engineering Design." Chapter 23 in *Design Management: A Handbook of Issues and Methods,* Oakley, M. (ed.). Oxford: Basil Blackwell, 1990.

The broad context of parameter analysis, even beyond its use as a methodology for conceptual design, is described in:

Li, Y. T., Jansson, D. G., and Cravalho, E. G. *Technological Innovation in Education and Industry.* New York: Van Nostrand Reinhold, 1980.

The inventor of the tiltmeter used in this chapter is Y. T. Li, the "father" of parameter analysis and a former colleague of D. G. Jansson. A full description of the tiltmeter is in the following patent record:

Li, Y. T. *Sensitive Tiltmeter.* U.S. Patent 3,997,976, 1976.

Some information on the dynamic aspects of the tiltmeter can be found in:

Redfield, R. C. and Krishnan, S. "Towards Automated Conceptual Design of Physical Dynamic Systems." *Journal of Engineering Design* 3, No. 3 (1992): 187–204.

5

Parameter Analysis Put to Work

The previous chapter presented the theory behind the parameter analysis methodology. This chapter introduces the preliminary step—*technology identification*—in which several core technologies are lined up to form the starting points for design. The three basic building blocks of parameter analysis—parameter identification, creative synthesis, and evaluation—are then described separately. Several guidelines and illustrative examples are provided to help the reader execute each of these steps efficiently. Finally, we focus on practical aspects of applying parameter analysis to conceptual design, and we outline some important characteristics of the process.

5.1 Parameter Analysis Process Overview

Figure 5.1 shows what this chapter is all about. Suppose that we have already finished a comprehensive need analysis in which we investigated the design task in depth and gained considerable knowledge about the problem domain, functions, and various constraints. We have also generated the design requirements, so now we are ready to begin with the actual design of the new artifact. When we are finished with the conceptual design stage (the dashed block in Fig. 5.1), we should have one or several well-developed conceptual designs. As you will soon learn, these conceptual designs are much more than general ideas of how to solve the design task. In fact, they are quite

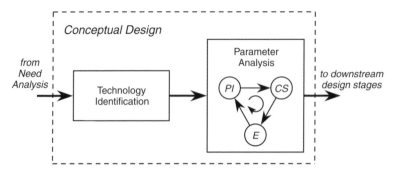

Figure 5.1 The conceptual design stage consists of technology identification and parameter analysis. (*PI*—Parameter Identification, *CS*—Creative Synthesis, *E*—Evaluation).

detailed configurations that have evolved through several cycles of parameter analysis.

The first block shown in the diagram is called *technology identi-fication*. It refers to the process of looking into possible technologies that can be used for the design task at hand, thus establishing several starting points, or initial conditions, for parameter analysis. Having selected a core technology, we can now begin the parameter analysis process. As discussed in the previous chapter, the parameter analysis methodology emphasizes the discovery of one or a few critical conceptual issues at a time, calls for implementing these concepts as configurations, and directs the designer to keep evaluating the evolving design to identify new, emerging dominant issues at the conceptual level. The process consists of going through cycles of the following distinct steps, which correspond to the above activities:

1. ***Parameter Identification (PI):*** Recognizing or discovering new important conceptual issues.
2. ***Creative Synthesis (CS):*** Generating a configuration, or hardware, that realizes the previous concept.
3. ***Evaluation (E):*** Assessing the last configuration.

Parameter analysis can be applied to any design task and at any level of innovation. It facilitates technological innovation, or technologi-cal breakthroughs, by assisting designers during the important conceptual design stage, where most opportunities for novel design solutions exist.

5.2 Technology Identification

Often, several core technologies, or physical principles, can be used in a particular design. This is especially true when designing sensors, but also with other design tasks, as discussed in the following example.

> **Technology Identification for a Robot Plow**
>
> Suppose we are assigned the task of designing an autonomous plow for agricultural use. What is needed is an automatic device—a sort of a farm tractor—that will plow fields with a minimum of human intervention. We perform a need analysis in which we quantify all the relevant considerations, such as field sizes and topographies, required force and speed of plowing, navigation and guidance, and cost. With our newly gained understanding and knowledge in the area of farming, as well as a list of design requirements, we proceed to the conceptual design stage.
>
> Realizing that the plowing operation itself is a well-established technique and that existing farm tractors do a good job of providing the necessary mobility and supporting platform for a plow, we decide to focus our effort on the navigation and guidance aspect of the design. *Technology identification*—the listing and assessment of possible means of guiding an autonomous vehicle in a field—may look like this:
>
> - Vision
> - Ultrasound
>
> These are technologies for sensors that can be mounted on the tractor and used to detect nearby objects. They seem more suitable for obstacle avoidance than for navigation.

- Buried line following

This technology is widely used in guiding autonomous vehicles on factory floors. A buried metal line, for example, could be detected and followed with the help of magnetic sensors. Three problems with this method are quickly recognized: The initial cost of installing the guiding lines might be very high; the plow would always follow the same path with no flexibility; and the plow might accidentally destroy the guiding line if the latter is not buried deep enough.

- Triangulation

The robot plow could find its location in the field at any time by measuring range, direction, or a combination thereof, in relation to some fixed targets. Comparing this location to a stored map of the field, an on-board navigation computer could guide the vehicle. The type of measurement can vary widely between radio signals, lasers, and more.

- Global positioning

Systems that rely on signals from satellites have recently become quite affordable. As with triangulation, the measured location of the vehicle should be compared with a pre-stored desired location. Whether the degree of accuracy obtainable with a commercial system would satisfy the requirement of this design is somewhat questionable, however.

This example is by no means a complete listing of pertinent technologies, but it demonstrates several advantages of carrying out technology identification. First, it offers additional insight into the problem domain. For example, if we had not realized it during need analysis, it now becomes clear that in addition to a navigation system for locating the tractor relative to its desired path, we may also need to provide a means of obstacle avoidance to increase the operating safety.

Second, technology identification offers us several possible initial conditions for parameter analysis. From the preceding list we select the technology that seems most likely to result in a successful design, and the first parameter, or concept, usually becomes: "use technology X to achieve task Y." For the robot plow example, X might be "triangulation" and Y, "navigation."

If more than one conceptual design is to be developed, as is often the case, different technologies identified at this stage can be used as the starting points to yield conceptually different results. And in case one parameter analysis attempt fails, the list of various candidate technologies can be revisited and another selection can be made for a second attempt.

5.3 Parameter Identification

The main concern during parameter identification is to generate a new and improved concept using scientific and engineering principles. This is done by examining all the information about the design task, the latest configurational solution to it, and the results of the latest evaluation step, and generating a new design parameter.

> *Design parameters are not design variables. They normally are not dimensions, material properties, or any other piece of configurational information, which in other contexts may be referred to as "parameters." Design parameters in the context of parameter analysis and this book are issues at the conceptual level. They may be the dominant physics governing a problem, a new insight into critical relationships between some characteristics, an analogy that helps shed new light on the design task, or an idea of what should be the next focus of the designer's attention.*

Dominating parameters may not be immediately apparent in a new situation but may have more influence on the outcome than those that are obvious. The following example illustrates some characteristics of parameters.

Critical Parameters in Electrical Fuse Design

Fuses are safety devices that prevent damage to an electrical circuit due to excessive current under overload and short-circuit conditions. The difference between these two conditions is the magnitude of the current involved, with short-circuit currents being orders of magnitude greater than overload currents.

It is quite obvious that a fuse must react faster than the event against which it is guarding. So a critical parameter in designing a fuse may initially be formulated as "make the time required for blowing the fuse shorter than the time it would take the excessive current to damage the circuit." This parameter is often quantified as a time–current graph, where the required time to blow is plotted for various currents.

Now suppose we are designing a fuse for short-circuit protection. To determine the critical parameters, we need to examine the underlying physical processes involved. The time required to blow a fuse depends on the heat generated in the fuse element (which is a function of the current flowing through the fuse element and its resistance), heat dissipated from the fuse element, and its material properties, such as melting temperature, specific heat, and latent heat. It seems that these energy considerations dominate the design and are sufficient to yield a satisfactory fuse.

A closer examination, however, will reveal that arcing may occur in the fuse during the blowing period. The arcing current continues even when there is a discontinuity in the fuse element. The critical parameter therefore becomes "control the arcing," which was not obvious from the initial energy considerations.

In the parameter identification stage, the designer spends considerable time and effort to properly define the current concept to be realized next. The extent of innovation achieved is largely governed by the designer's ability to recognize crucial relationships that dictate possible solution paradigms. An innovative design often results from the designer identifying parameters—concepts and ideas—that make some relationships irrelevant and break away from current solution paradigms. This is illustrated by the next example.

Neutral Buoyancy: A Common Parameter in Submarine Design

Designing a submerged structure to be neutrally buoyant is a widely used concept. An obvious example is a ship, but the gates of the Panama Canal were also designed to be neutrally buoyant to relieve the hinges from carrying the weight of the gate. The same concept, or parameter, dominates the design of contemporary submarines.

Submarines rely on stealth for their functioning and survival, so a small submarine that can go deep is desirable. However, deeper-going submarines experience higher hydrostatic pressures and require thicker, heavier hulls. A crucial consideration in submarine design is therefore to maintain neutral buoyancy by increasing the size of the submarine to where the weight gain due to the larger hull thickness is balanced by the weight of the additional water volume displaced. The result is a direct relationship between the operating depth and the size of the submarine.

The key to radical innovation in submarine design lies in discovering such fundamental relationships and identifying concepts that might make them irrelevant. For instance, submersibles that use hydrofoil or vector propulsion to raise and lower themselves can have some advantages in certain naval activities. And in this case, because the submarine does not need to be neutrally buoyant, its size is almost independent of the operating depth. An analogous paradigm shift occurred when air transportation moved from the "lighter-than-air" airships to "heavier-than-air" airplanes.

Parameter identification includes two key activities: *simplification* and *transformation*. Simplification of a design task is aimed at making it less complex, and it usually involves the following activities:

- **Stripping the task of less important considerations:** "Less important" is a temporal term that depends on the exact stage in the conceptual design process. Issues that are less important early in the process may later be identified as the critical parameters or may only be tackled during embodiment or detail design.

  For example, the double-pendulum tiltmeter design described in Chapter 4 would not be a viable solution without devising a configuration for nearly frictionless hinges and an effective means of measuring angles or lateral displacements. However, these issues are only secondary at the beginning of the design process, when the more critical aspects are the view of a pendulum as a spring and obtaining a small stiffness by subtracting two large stiffnesses. Moreover, the hinges and measurement device cannot be identified as critical parameters at the outset of the design process because they only emerge later, in connection with the particular configuration developed.

- **Focusing on make-or-break issues:** While several issues confront the designer at each step of the conceptual design process, he or she must be able to identify those aspects of the task that, if not done right, would jeopardize the whole design. Designers focus their efforts on solving these issues first.

  Consider, for example, the robot plow discussed in the previous section. The actual plowing operation (i.e., disturbing the soil with the plow blade), propelling, braking and steering a vehicle, and designing a tractor-like vehicle to traverse various terrain, are all secondary issues that we believe we know how to handle. What turns out to be of primary significance at the beginning of this design process are the issues of navigation and accuracy: The autonomous vehicle has to know where it is and where it is supposed to be, and this must be done with certain accuracy which is defined by the requirements. From our understanding of the state-of-the-art of technology, we judge that the navigation and accuracy are the make-or-break issues in this design task.

- *Developing simple models to understand governing relationships:* It is often necessary to develop simple models that capture some essential relationships. These models can provide deeper understanding of the underlying scientific principles and identification of potential conflicts or inconsistencies that need to be resolved during the design process.

Understanding Essential Relationships in Car Headlights

The design of car headlights poses an interesting challenge. Two key requirements for headlights are: provide visibility for night driving and avoid the blinding of oncoming traffic. In the traditional headlight design, these two requirements can be satisfied by varying the brightness of the light. As shown in the relationship model of Fig. 5.2, increasing the brightness positively affects visibility and negatively affects blinding. Thus, the model accentuates a key conflict in headlight design.

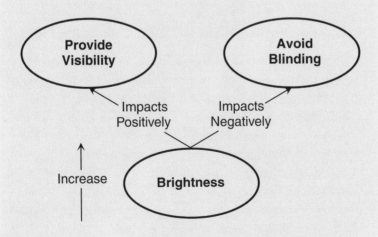

Figure 5.2 A simple relationship model for car headlight design.

It is up to the designer to continually judge what are the critical parameters at any given point and to differentiate them from the less important issues. To a large extent, the quality of the whole conceptual design process depends on this ability of the designer. By focusing only on the most dominant parameters, the overall design task is simplified, thus increasing the chances of discovering or changing the fundamental principles and concepts of the design.

The second important activity during parameter identification is transformation of design parameters. This is the thought process of gaining new understanding of the task, and it often takes place when one concept is abandoned in favor of a new solution approach. Searching for a new conceptual direction, the designer identifies new design parameters that may lead to the generation of new ideas. Design parameters are often transformed as a result of one or more of the following activities:

- *Rephrasing the task statement:* Restating the design task in a way that emphasizes the important issues and postpones the consideration of secondary parameters. This often involves replacing some key words with more appropriate ones and allows the designer to view the design task differently.

Rephrasing the Task Statement for a CT-Scanning Fixture

A large defense manufacturer needed a universal fixture for holding assemblies of various sizes and shapes at any desired orientation for quality control of the inner components by computed tomography (CT) scanning. The original task statement—"design a fixture to hold objects at any orientation"—led the designers to attempt a structure that would securely hold test objects at any orientation given to them by the human operator before placing them on the fixture in the CT scanner (Fig. 5.3a). After much struggle with this design task, a breakthrough took place when it was realized that the object could first be clamped in a fixture in some easy-to-clamp orientation, and then

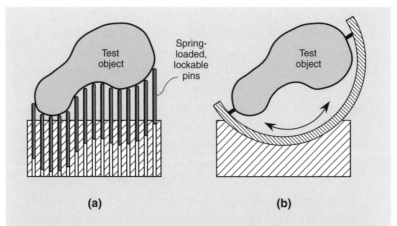

Figure 5.3 (a) A schematic of a fixture to hold a test object at any orientation. The pins are spring-loaded and conform to the shape of the object before being locked in place. (b) A simpler solution in which the object is first clamped and then oriented with the movable part of the fixture.

the object plus fixture could be given the desired orientation as one unit. The modified task statement—"design a fixture to hold and orient objects"—soon resulted in a successful solution, shown schematically in Fig. 5.3b.

- *Explaining the cause of a problem in a different way:* This activity is strongly related to the need to understand underlying physics as a precondition to successful design.

Understanding the Underlying Physics of Clear-Ice Manufacturing

Ice served on airplanes and expensive restaurants is clear, whereas the ice cubes frozen in refrigerators are hazy or cloudy. At a superficial level, the haziness is attributed to the presence of dissolved air in the water. During solidification, the air separates from the water and forms small air bubbles. An obvious solution would seemingly be to eliminate dissolved air in the water. After some design effort, however, this task turns out to be very difficult, forcing the designer to further investigate the root cause of haziness in the ice.

Water at room temperature contains a certain amount of dissolved air. As the water is cooled, ice crystals start to form on the outside surfaces and proceed to grow inwards when the temperature is reduced further. Because the solubility of air in ice is very small, these crystals—known as primary ice—release dissolved air to the liquid water, thus increasing the concentration of dissolved air in the liquid. Then, at the eutectic point, the remaining water (with the dissolved air) freezes to form secondary ice. During this reaction, the air separates out of the ice into small air bubbles, which produce the undesired cloudiness in ordinary ice. Thus, the primary ice formed in the early stages tends to be clear, whereas the secondary ice is cloudy. When designing a device to produce clear ice, the critical parameter becomes "prevent the liquid solution from reaching the eutectic point." This insight is used in building most of the commercial clear-ice machines.

- *"Stepping back":* Reexamining the design task from a different perspective.

Different Perspectives on Controlling Industrial Robots

The accuracy of industrial robots relies heavily on the rigidity of their structure and elimination of backlash in the joints. The position of the end-effector of a robot arm is determined by sensing displacements at the joints. Any deflection of the arm or backlash in the mechanism would create a discrepancy between the calculated and actual locations. Robot designers always attempt to use stiffer structures and tighter tolerances; however, a totally different perspective on this issue is also possible: If it is the location of the end-effector that needs to be controlled, then why not sense this location directly? Suppose some noncontact sensor is used to measure the location of the robot arm tip. One possibility is to follow the end-effector with laser beams to determine its location. In this case, the mechanical structure of the robot can be made much less rigid and expensive. The resulting lower weight and inertia will also improve dynamic performance through higher speed and acceleration.

- ▪ *Parameter substitution:* Replacing one design parameter with another, or with a combination of other parameters, which is more fundamental.

Parameter Substitution in Total Hip Replacement

Artificial hip reconstruction surgery is necessary in the case of hip joint failure. About 170,000 Total Hip Replacement (THR) surgeries are performed each year in the United States alone. In the THR surgery, the surgeon replaces the head of the femur and the socket part of the pelvis bone. The femur portion of the surgery involves removing the head of the femur, drilling a deep hole into the medullary canal, and inserting the stem of the prosthesis into it. Historical data suggest that 10% of THRs fail in ten years and 20% in twenty years.

An important issue faced by the designers of the prosthesis is "stress shielding," the phenomenon of bone loss and reduction in bone thickness due to the decrease in the load on the bone. The stiffness of metals is much greater than that of the bone. For instance, Young's modulus for titanium is 150×10^5 psi, while that of bone is 6×10^5 psi. A solid implant that conforms to the drilled hole in the femur has a much greater stiffness than the bone it replaces. When load is applied, the stiff metallic stem deforms very little and therefore carries most of the load. The bone, on the other hand, experiences very little load due to its very small deformation. This quick look at the problem suggests that the critical design parameter is "find a material whose elastic modulus is identical to that of the bone."

It turns out that such matching of elastic moduli is very difficult. Instead, a new key parameter is stated as follows: "match the *stiffness* of the implant with that of the bone." By replacing "elastic modulus" with "stiffness," designers can conceive different ideas because stiffness depends on both the material and geometry. THR designers employ various structural principles, such as the use of porous materials, to match the stiffnesses. Note that *matching* of some property is a common and fundamental design parameter that plays a role almost any time individual components are made to work together.

- *Analogical thinking:* The thought process of recalling a familiar configuration and adapting it to a new situation. The configurations are analogous because they realize the same design concept, or parameter. For example, airbags have been recalled from the automotive field and adapted for cushioning the landing impact on Mars. Analogies can also occur at a deeper, more conceptual level. For instance, stress concentrations are often reduced by understanding the stress field using flow analogies. An analogy to electrical fuses, based on their need to react faster than the event against which they guard, may be useful in automotive airbag design. The role of analogies is demonstrated in the following example.

Analogies with Aluminum Foil

Aluminum foil used for cooking has a shiny side and a dull side. For heating food, the shiny side should be placed toward the food, so it reflects radiant heat back to the food while the dull side absorbs the external heat. This configuration thus creates a bias in the radiation heat transfer. The same concept has been used in the design of chocolate wrappers and car sun shades, where the glossy outside reflects heat away to keep the chocolate or the interior of the car cool.

The design of space suits poses an interesting challenge. Without an appropriate space suit, an astronaut's body heats up due to solar radiation when facing the sun and loses heat when in the shade. The temperature fluctuations could reach ±400°F. In this particular application, the critical parameter is "block the solar radiant heat from entering the space suit while reflecting the body heat back." In other words, the design must integrate the features of chocolate wrappers to keep the external heat away with the features of cooking foil to retain internal heat. NASA uses a multilayered, double-shiny foil to achieve this task. The same principle is used for thermal shielding of spacecraft components and for building insulation.

Every cycle of parameter analysis (Fig. 5.1) will always have a parameter identification step. Even when it seems that a configuration can be changed by mentally staying in "configuration space," this is never really the case. When the designer examines a configuration and realizes that something needs to be modified, he or she has actually performed an *evaluation* and then identified a new critical issue—a parameter—that now becomes the conceptual basis for the configurational change.

Although good identification of dominant parameters is the key to successful conceptual design, not every single parameter identified in the course of developing a design must be a "breakthrough parameter." Some parameters are in fact quite mundane. Nevertheless, their presence as a component of the parameter analysis process provides the necessary conceptual "justification" for any physical aspect of the evolving design.

Theoretically, the most significant conceptual issues—the dominating parameters—are identified early in the design process, and as we proceed downstream we encounter and handle the more routine issues. However, design is rarely such a linear process, and it often takes a large effort and many cycles of parameter analysis before the real dominant issues and concepts emerge. In this sense, the intellectual effort during parameter analysis, including failed attempts and taking wrong paths, can be regarded as a stimulant and contributor to the eventual success in identifying the real issues of the design task.

5.4 Creative Synthesis

The new design concept, stated in previously identified parameters, becomes the focal point for generating design configurations during the *creative synthesis* step. A design artifact is proposed here to solve, satisfy, or embody the conceptual parameters. Sketches and "back-of-the-envelope" calculations are used extensively during this step to provide enough detail about the design and its operation.

A configuration will usually be represented as a sketch and should always be quantified. By this we mean that the sketch usually contains some dimensions, perhaps material selections, and other details. This will ensure that analysis can be applied and the configuration can be evaluated for its extent of meeting the design requirements (which, of course, are quantitative, too).

Creative synthesis may be a highly divergent step in the sense that many configurations can be proposed at any given time to satisfy the same set of parameters. In this case, the next evaluation step will be used to select among these configurations. Alternatively, if only one configuration is proposed to embody the concept, then the next evaluation will determine its viability, and, according to the results, another parameter may be identified, to be followed by a different or modified configuration.

Skyscraper Configurations to Resist Wind Loads

Fazlur Khan designed two of the tallest buildings in the world—the John Hancock Center and the Sears Tower, both in Chicago. In the design of high-rise buildings, one of the design challenges is handling lateral wind loads. Gustave Eiffel addressed this issue in the design of the Eiffel Tower by tapering it to reduce the wind loads on the structure. In the John Hancock Building (schematic shown in Fig. 5.4a), Khan used a gentle taper to

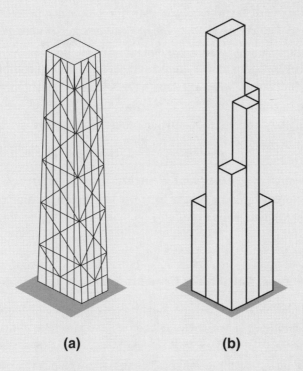

(a)   **(b)**

Figure 5.4 Two skyscraper configurations to resist lateral wind loads: (a) The X-braced tube of the John Hancock Center and (b) the bundled-tube of the Sears Tower.

reduce the wind loads and 18-story tall exterior steel diagonals—the famous X-braces—to carry this load. Thus, the configuration of the building is a tube stiffened by the floors. Later, Khan designed the Sears Tower (Fig. 5.4b) using another innovative tubular system, known as the "bundled-tube" approach. Nine steel-framed tubes, each 75 feet square, are interlocked to form a massive column whose combined thickness can easily resist wind loads. Both of these "rigid-tube" configurations eliminated much of the internal bracing that could eat up costly floor space.

As the word "synthesis" suggests, the creative synthesis step is concerned with creatively integrating different constituent elements, or building blocks, into a unified design. In the case of system design, the building blocks are subsystems and system components. For product design, they may be the machine and structural components. In component design, the constituent elements are materials, geometry, and manufacturing processes. Creative synthesis is therefore about

- *Identifying constituent elements:* The designer can choose different building blocks to create the design, as illustrated by the following example.

Identifying Constituent Elements in Explosives Design

The effectiveness and safety of explosives are measured in terms of power and sensitivity to detonation. The power rating compares the volume of gases produced by the explosive to that of picric acid, which is given a rating of 100. The higher this number, the more powerful is the explosive. Sensitivity to detonation is the tendency to explode when a force is applied. Picric acid, with a rating of 100, is the standard of comparison for sensitivity, too. Explosives with a sensitivity rating less than 100 are considered unstable and unsafe.

Nitroglycerine is a powerful explosive that releases 12,000 times its own volume in gases when exploding. It has a high power rating of 160 but also a sensitivity rating of 13. This means that it is extremely sensitive and may explode even by "a mere shake." No one had been able to harness the potential of nitroglycerine because its sensitivity made the handling and transportation extremely dangerous. To allow practical use of nitroglycerine, its sensitivity would have to be suppressed while retaining its explosive power

Alfred Nobel succeeded in solving the problem by identifying another constituent element—kieselguhr, a very porous diatomaceous earth. In dynamite, Nobel used kieselguhr to absorb nitroglycerine without reacting with it, thus retaining the explosive power while reducing sensitivity. The resulting plastic explosive was powerful and safe to handle. Dynamite has since been used as a safe and effective explosive for military, construction, and demolition applications.

- *Integrating constituent elements:* The designer can combine the building blocks in different arrangements, both spatial and temporal. Even though the design may use the same components, the level of innovation and the overall performance depend on their arrangement. For instance, the parts of the Sony Walkman were designed to fit together from one direction, allowing the use of efficient robotic assembly. The following examples describe how constituent elements (geometry, material, and manufacturing process) are integrated in a single component.

Foil Manufacturing

The functional need to create a bias in radiation heat transfer can be addressed by a foil with shiny and dull sides, as described in the previous section. To manufacture this unique combination of features, designers conceived the process of pack rolling, where two foils are rolled together. The sides of the foils facing the rollers become glossy, and the sides facing each other become dull. The material—aluminum—is an excellent choice because it is very ductile and can be formed into very thin sheets (5–6 mil). Thus, the foil configuration integrates geometry (two different sides), manufacturing process (pack rolling), and material (aluminum).

The Shape of Manhole Covers

The circular shape of manhole covers addresses several design issues. A circular cover will not accidentally fall into its hole, whereas all other shapes might, if given the right orientation. A circular cover is easy to move by rolling it. It is also stronger than other shapes because the center point is equidistant from the periphery, where it is supported. Such synergistic integration during creative synthesis is one of the keys to innovation.

Why Do Coins Have Rims?

Two main considerations in the design of coins are "allow stackability" and "prevent wear of the faces." These two issues are addressed by incorporating a rim at the edge of every coin configuration. The rim is higher than the face, so coins can be stacked easily. Wear is also prevented because the coins do not touch each other at the face.

5.5 Evaluation

This third step in parameter analysis serves two purposes: assessing the design performance and providing additional design information. The latest configuration is examined in this step in light of the conceptual goal and overall design requirements. Weaknesses are highlighted to help identify the next design parameter. Three distinct evaluation methods are: (1) comparison of the configuration to the original requirements; (2) identification of weaknesses; and (3) comparison between several configurations. These evaluation methods are demonstrated by the following examples.

Evaluation by Comparison to the Requirements: Battery-Powered Cars

California's Zero Emission Vehicle (ZEV) regulations stipulate how many cars sold in California should be ZEVs in the coming years. Recognizing that other states and countries would soon follow a similar course, the auto industry is focusing on developing ZEVs in general and electric battery-powered cars in particular. One of the performance criteria in using battery technology is specific energy—the amount of energy that can be stored per unit mass for a specified rate of discharge. Gasoline's specific energy is 12,000 Wh/kg, and this figure establishes a convenient yardstick in the design requirements for electric vehicles.

An evaluation of existing battery technology reveals that the theoretical maximum specific energy for lead-acid batteries is about 170 Wh/kg, with current technology being around 35–40 Wh/kg. Current lithium-ion battery technology can provide about 120 Wh/kg. Table 5.1 summarizes the characteristics of three current battery technologies.

Table 5.1 Comparison among some automotive battery technologies.

	Lead-Acid	Nickel-Metal Hydride (NiMH)	Lithium Ion (Li-Ion)
Advantages	The first commercially available rechargeable battery. Cheap to mass produce. Low maintenance.	Clean, reliable, and maintenance-free. Range of up to 150 miles per charge. Expected to last the life of the vehicle.	Very light-weight.
Disadvantages	Moderate range. Cold weather reduces storage capacity.	More expensive than lead-acid batteries.	More expensive than NiMH batteries.

Evaluation by Identifying Weaknesses:
Fuel-Cell Powered Cars

Fuel cells are an alternative to batteries for powering cars, and automakers are experimenting with them. A fuel cell combines hydrogen and oxygen into water, as shown in the diagram of Fig. 5.5. In the process, it produces heat and electricity. The catalyst helps hydrogen to separate into a proton and an electron. The electrons constitute the electric current, while the protons pass through the electrolyte and combine with oxygen to form water.

The hydrogen can be either stored on-board in a compressed state, in an absorbed state in metal hydride, or obtained from methanol. Fuel cells that use methanol have a

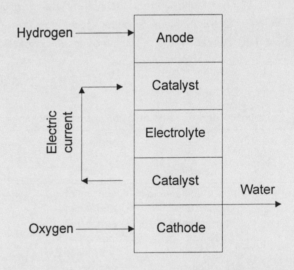

Figure 5.5 Hydrogen and oxygen combine in a fuel cell to produce electricity and water.

high specific energy of 2000 Wh/kg. Fuel cells with Proton Exchange Membrane (PEM) electrolytes are being developed for cars. They are very expensive because they use platinum. New technological breakthroughs, which will reduce the amount of platinum needed, are required before this technology can be used commercially.

Evaluation by Comparing Several Configurations: Hybrid-Powered Cars

As the name suggests, a hybrid vehicle has two energy sources: an internal combustion engine and an electric battery. These can be connected in parallel or series, as shown in Fig. 5.6.

In the parallel configuration, the power from the engine is combined with the power from the electric motor. Maximum

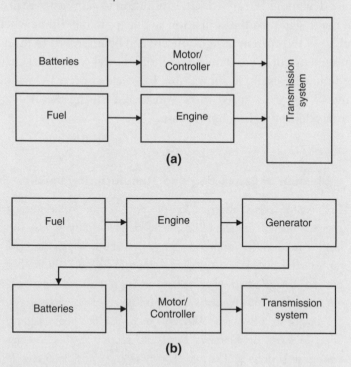

Figure 5.6 (a) A parallel arrangement and (b) a series arrangement for the power train of a hybrid vehicle.

power is therefore high, and a generator is not needed. However, this arrangement requires the difficult job of joining the torque from two power sources. On the other hand, the series configuration reduces emissions by eliminating the idling of the engine and greatly simplifies the transmission system. Both these arrangements minimize energy losses and emissions, and reduce the fuel consumption by 50% for the same power output.

To determine how well a design measures up to its expected performance, it is necessary to evaluate the configuration quantitatively. This requires that the configuration contains enough detail, such as materials information and dimensions, and that approximate calculations be carried out. In addition to gaining more insight into the design task, the evaluation step helps to determine whether (and how) the current design concept can be improved to meet the requirements, or there should be a more radical conceptual change. The following example shows how life-cycle concerns—manufacturing in this case—need to be considered during the evaluation stage in addition to functional issues.

Evaluation of Connecting-Rod Manufacturing Details

Connecting rods are forged to their final shape. In the forging process, a small amount of flash extends at the parting line. This flash is removed by grinding. The two methods of grinding are represented by the two grinding wheels in Fig. 5.7. In the first method, the grinding wheel rotates in the plane of the connecting rod, whereas in the second method the grinding wheel is perpendicular to this plane. Although both methods seem equally good in terms of removing flash, the second method creates transverse surface cracks, which may lead to what is known in fracture mechanics as Failure Mode I, which is very dangerous.

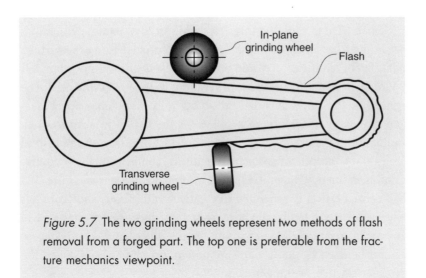

Figure 5.7 The two grinding wheels represent two methods of flash removal from a forged part. The top one is preferable from the fracture mechanics viewpoint.

The evaluation is by no means a mere filtering mechanism. The main purpose is not finding fault but, rather, constructive criticism. A well-balanced observation of both good and bad aspects of the design is crucial to pointing out possible venues of improvement in the next design cycle. In terms of the designer's thought process, the evaluation step belongs in both configuration and concept spaces. It involves an assessment of the latest embodiment of the design (clearly, a member of configuration space) in light of the design parameters (the conceptual goals and critical issues that belong in concept space).

5.6 Parameter Analysis Revisited

The solution to a design task consists of two components: a physical configuration and a concept, that is, the operational principles behind the configuration. Using parameter analysis, we find that the design solution evolves through multiple cycles of triplets: *parameter identification* to generate a concept, *creative synthesis* to embody the concept in "virtual hardware," and *evaluation* to show the direction for the next cycle.

If we were to rank the three repeating steps in parameter analysis according to significance in the overall conceptual design process, then parameter identification would turn out to be the most important, followed by evaluation, and creative synthesis as the least influential. The reason for this ranking is that the ability to conceptualize novel ideas and new ways of looking at problems is the key to true innovation. Evaluation is the driving force behind the identification of new critical issues and concepts, while creative synthesis, resulting in a new configuration, is "merely" an embodiment of previously generated ideas. If a bad configuration is generated in creative synthesis, its shortcomings will easily be pointed by the next evaluation, and a better concept will emerge in parameter identification. On the other hand, a bad concept cannot be fixed by good hardware design, no matter how skilled the designer is while working in configuration space.

The cyclic character of parameter analysis, that is, repeatedly refining the design through cycles of analysis and synthesis, combines two important characteristics of any design process: divergent and convergent thinking. The mental processes in concept space (see Fig. 4.9), namely, parameter identification and part of evaluation, are convergent in the sense that they focus the design progression by narrowing down the choices. On the other hand, activities in configuration space—creative synthesis and the other part of evaluation—tend to be more divergent. When we are at the parameter identification step, we concentrate on one or a few critical issues, and we set aside the rest. Then, in creative synthesis, we may examine many ways of realizing that idea, but we use the evaluation step to perform an abstraction that focuses us again on the most critical parameters. This repeated combination of divergent and convergent thinking is depicted in Fig. 5.8.

While Fig. 5.8 may suggest that parameter analysis is a very linear process, with a distinct beginning and ending, this is usually not the case. Recall that the process generally begins by selecting a single technology from several possible ones recognized during the technology identification stage and by using that technology as a starting point. It may well happen that a long parameter analysis process will culminate in a dead-end, meaning that at some point it becomes apparent that the fundamental concepts are erroneous or that the

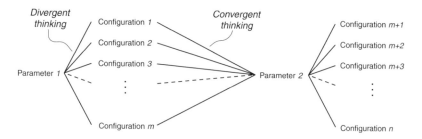

Figure 5.8 Going through the steps of parameter analysis corresponds to alternating between divergent and convergent thinking modes.

concepts are good but their technological realization is infeasible. A characteristic of a good designer is his or her ability to recognize such situations and instead of spending more effort and resources, backtrack and start over. The designer may use a different technology for the second attempt at conceptual design, or a less radical approach by choosing a different configuration among several possibilities at any earlier stage of parameter analysis. The many paths from the initial technology identification to final solutions are shown schematically in Fig. 5.9.

Suppose point A is the starting point for conceptual design. Several technologies are identified here as feasible approaches to the design task. The designer chooses one of these technologies and initiates the parameter analysis process along path *I* in Fig. 5.9. (Parameter analysis paths are shown as curly lines to symbolize the many cycles of parameter identification, creative synthesis, and evaluation, which constitute each path.) The process reaches a dead-end at point B, at which the designer decides to attempt a different configuration instead of the one used earlier in the process, say, at point C. The designer then discards everything that took place between points C and B, and moves in a new path (*II*) from C to D, where a satisfactory solution is reached.

Path *III* in Fig. 5.9 shows another situation. A dead-end is reached at E, and the designer decides to try a totally different approach, or technology, for the conceptual design. Backtracking to the starting point (A) takes place, another technology is chosen, and a new path (*IV*) is taken until a new solution is created at point F.

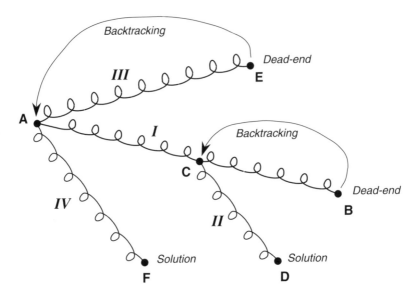

Figure 5.9 A schematic of the branching character of parameter analysis.

The decisions described here, namely, to backtrack partially but stay on the same path versus complete backtracking and starting over, are not easy to make. On the one hand, a good concept should not be abandoned too early in the face of difficulty; rather, such difficulties are often the driving forces that lead to valuable new insights in concept space. On the other hand, there is not much point in insisting on developing a fundamentally bad design, especially when realistic time constraints and limited resources are present. In the end, it is up to the skillful designer to make the correct decision. The following example demonstrates some of these issues.

Backtracking in Magnetic Seal Design

Centrifugal pumps used with chemicals usually employ a dynamic seal between the shaft and the casing. This seal is often a major source of leakage. In an attempt to overcome this problem, engineers came up with a novel idea: transmit the motor torque by a set of permanent magnets placed across a physical barrier, thus eliminating the dynamic seal altogether. Let us follow several steps in the parameter analysis process, starting with this idea:

PI: Use magnets across a physical barrier to transmit torque while sealing.

CS: This idea is realized in "magnetic drives," shown schematically in Fig. 5.10. Permanent magnets are installed on both the motor and the pump shafts and are separated by a static barrier. The barrier needs to resist corrosion from the pumped chemicals and withstand high tensile stresses (from the fluid pressure), so it is made of steel.

E: A major side effect of metallic barriers is the generation of eddy currents due to the relative motion between the conducting barrier and the magnets, and the electrical resistivity of the barrier material. This causes hysteresis losses that could be greater than 50% of the total power transmitted.

PI: Eliminate eddy current losses by using a nonmetallic barrier.

CS: Use a ceramic barrier.

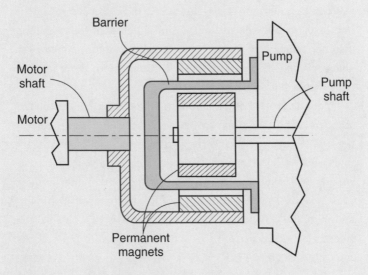

Figure 5.10 Schematic of a typical magnetic drive.

E: The fluid pressure in the pump causes tensile stresses (circumferential and longitudinal) in the barrier, and ceramics are not good at carrying tensile loads.

PI: There are at least two choices now. One is to pursue the ceramic barrier solution further, which may involve introducing fiber reinforcement to carry tensile loads. The second is to look at alternative methods for reducing eddy currents. The following insights can be gained from studying the underlying physics:

Eddy-current losses are

- directly proportional to the square of the speed;
- directly proportional to the square of the radius of the drive;
- directly proportional to the square of the flux density; and
- inversely proportional to the resistivity of the barrier material.

To improve cost effectiveness, the amount of magnetic material required can be reduced by decreasing the distance between the two magnets. However, the increase in flux density between the magnets will increase the eddy current losses and result in a lower efficiency. The diameter of the drive can also be reduced. However, both approaches can improve the efficiency only to a certain level. Fundamentally, the eddy currents would still persist.

We can look at other situations where designers deal with eddy current losses, such as electric transformers and motors. The transformer core and the motor rotor use laminar construction to minimize eddy currents. Eddy currents decrease linearly with the number of layers because the flux driving them in each layer is reduced. The I^2R losses decrease with the square of the number of layers. If the barrier is made of two pieces, the eddy current losses will decrease to 25%. Four pieces will decrease the losses to 6.25%. The new parameter is therefore: "decrease the eddy currents by using a layered construction."

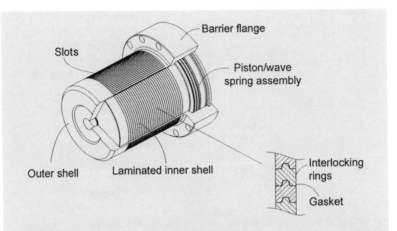

Figure 5.11 A seal design that significantly reduces eddy currents while adequately carrying the pressure-induced loads. (Courtesy of Nova Magnetics Limited.)

CS: Nova Magnetics Limited, a Canadian company, came up with the innovative solution shown in Fig. 5.11, which uses laminar construction. The inner shell consists of rings that enclose the fluid and carry the pressure-induced circumferential load. Because the thickness of each ring is small, the eddy current losses in this design are negligible. High-pressure gaskets are provided between the rings to prevent leakage of chemicals. The outer shell carries the longitudinal load due to the fluid pressure, and the rings and gaskets are live-loaded and held in place by a large wave spring and retaining ring. This outer shell is also slotted longitudinally to help minimize eddy current losses in it. In addition, the magnets are assembled in such a way as to help reduce eddy current losses. The design uses an alternating magnet pole configuration down the length of the coupling. This causes the direction of the eddy current generated by each magnet set to reverse when compared to the adjacent magnet set. Following Kirchhoff's Laws, this effectively cancels the movement of eddy currents from one magnet area to the other, thereby reducing resistive losses.

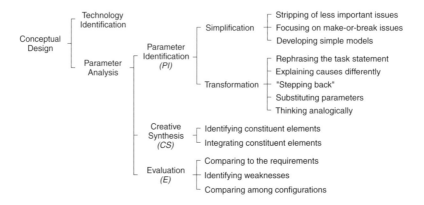

Figure 5.12 Summary of the activities in conceptual design.

5.7 Discussion and Summary

Conceptual design was shown to be composed of two stages: technology identification and parameter analysis. Core technical concepts that may be used for the design task at hand are listed first, and the one with the highest success potential, in the designer's opinion, is chosen as a starting point for parameter analysis. Parameter analysis consists of repeatedly cycling through three distinct steps: parameter identification, where concepts and ideas are generated; creative synthesis, where a configuration is devised to embody and realize conceptual insights; and evaluation, where the design is assessed constructively. Many activities and mental mechanisms that are associated with these steps were described in the chapter and are summarized in Fig. 5.12.

A basic question associated with parameter analysis is: how do we know that we are done? In other words, at what point should parameter analysis stop and the current configuration be declared a finished conceptual design? The answer is quite simple to state: parameter analysis should continue as long as there is doubt about the success of the configuration in meeting all the design requirements. Parameter analysis can be terminated only when no unresolved issues remain and all foreseeable questions regarding the operation of the design have been answered. The evaluation step is the point at which such determination is made, so the last step in parameter analysis is usually evaluation. This guideline implies that conceptual designs developed by parameter analysis are much more detailed than what many practices and textbooks call "conceptual design." The latter usually consists of a very general idea or just a rough sketch, as opposed to the well-developed configuration of the former.

5.8 Thought Questions

1. Suppose that your task is to design paper currency for use by blind people. Perform technology identification, listing as many core technologies as possible to help the visually impaired sense the value of bills. Critically assess each technology for its potential to develop into a viable solution. Keep in mind economical and usage constraints. For example, making notes of different sizes may be expensive, and the effect of impregnating them with different odors may not last very long or may be ineffective due to contamination by other odors.

2. Perform technology identification for the task of providing lighting in undersea applications. Remotely operated vehicles used in offshore oil drilling need the lighting to provide clear video images. Manned deep sea submarines used in rescue missions need to provide adequate visibility to their crew.

3. Identify the underlying physics, which may be the critical parameters, in the design of the following devices:
   (a) Different types of positive displacement pumps
   (b) Heat pipes
   (c) Piezoelectric actuators

4. Two important performance criteria with ink-jet printers are resolution and speed. Speed is usually measured as pages per minute, or ppm. Investigate the physics of injecting small ink droplets to understand what might limit the maximum operating speed of single-head printers.

5. Several generic types of bridges are in common use. Study the configurations of beam, arch, suspension, cantilever, and cable-stayed bridges. How does each configuration handle vertical loads, such as the weight of cars or trains crossing the bridge?

6. Most coins are round. Write some of the design requirements for coins, and evaluate round, square, and hexagonal coins against the requirements and against each other.

7. Some applications require a spring that applies a small but uniform force over a large displacement. A good example is the spring supporting the tailgate lid in hatchback cars. Identify different configurations that can be used for this spring design.

8. Different vacuum cleaning technologies are currently in use. Identify the technologies and discuss the advantages and disadvantages of each one.

6

Conceptual Design Case Study: HVAC Airflow Sensor

As the first detailed case study of conceptual design using the parameter analysis methodology, this chapter presents a relatively simple design example. Only a partial need analysis, highlighting the most important issues, is given here. The discussion under the value category uncovers the importance of keeping the cost down, and this issue also proves to be the dominant parameter in the development of the concept. This case study demonstrates the sequential evolution of the design throughout the parameter analysis process.

6.1 The Initial Need

Heating, ventilating, and air conditioning (HVAC) systems in buildings consist of a heating source, a cooling source, a fan to force airflow, and a closed-loop temperature control system. The last-named component usually includes a computer to compare actual temperature readings to set, or desired, values and activate the system accordingly. Such a system can handle the temperature well, but it does not account for the "freshness" of the indoor air. If conditions happen to be such that the temperature does not need to be actively changed by the system for a relatively long period of time, the air may become uncomfortably stale. The need is therefore to design a sensor to measure the airflow in HVAC ducts to allow the control system to monitor the number of air-volume exchanges in the rooms.

6.2 Abbreviated Need Analysis

Performance

The sensor needs to tell the computer how much volume of air per unit time passes through the duct. An obvious way would be to measure the flow velocity in the duct and to let the computer multiply the reading by the cross-sectional area of the duct to produce a volume-per-time reading. Alternatively, the sensor could be calibrated to suit the intended duct during manufacture or installation. In the latter case, readings of volume-per-time can be provided directly. The question then is, what range of velocities should the sensor be capable of measuring? The answer can be found by looking up information for typical HVAC systems, revealing that flow velocities of 50 to 1000 ft/min cover a wide range of duct sizes.

Every sensor has some properties associated with its measurement. Accuracy, defined as the deviation of the reading from the actual quantity being measured, is the first that comes to mind. For the HVAC sensor, it does not seem that a very high absolute accuracy is needed in general. Without presenting a detailed discussion of error analysis (this can be found in many books on measurement), we set the desired sensor accuracy to ±3% of full scale, or ±30 ft/min.

Linearity is another property that may be relevant. Linearity was important in the precomputer days, when Laplace Transforms were used extensively in the design of control systems. However, a computer can easily accommodate a nonlinear output from a sensor, so the potentially demanding requirement of linearity should not be posed for the HVAC sensor.

Next, we examine the environment in which the sensor functions. HVAC ducts typically accumulate considerable amounts of dirt and dust, and the functioning of the sensor should not deteriorate noticeably over time. It is difficult to quantify this item with regard to type and amount of contamination; however, the designer should be aware of this issue. If the sensor is relatively sensitive to the effects of contamination, more frequent maintenance will be required. This matter is also addressed in the next section.

Value

In order to establish target figures for manufacturing, installation, operating, and maintenance costs of the sensors, we need to fully understand the savings that may be realized by using the airflow measurement scheme. Considering a temperature-control-only HVAC system may be misleading because such a system does not address the question of air freshness at all. The real contender is an open-loop system, which provides a change of air volume every preset time. This overly simplistic method would probably result in too many air changes with the accompanied waste of energy. (Energy is required to force the new volume of air through the system, and even more energy may be consumed in bringing that volume of air to the desired temperature.)

A study of many office buildings, which typically have hundreds of sensor-mounting locations, would show that the sensors must be very inexpensive, selling for about $10 to $15 each. Manufactured cost therefore must only be at most a few dollars. Implied by this cost ceiling is the realization that electronics would probably have to be shared: instead of a signal-processing unit in each sensor, the signals would be sent to a central computer. Keeping the overall cost down also entails easy installation and low maintenance requirements. Climbing into HVAC ducts routinely to clean sensors should be avoided, so a requirement of at least five years of maintenance-free operation is generated.

Size and Safety

We shall not elaborate on these categories because the needs are quite obvious. The sensor should fit within existing ducts without disturbing the normal operation of the system, and it should comply with all building safety codes. The sensor should not in any way reduce the quality of air circulating in the system. Moreover, examining layout drawings of typical HVAC systems shows that air is usually fed to rooms and offices through horizontal or vertical ducts, so the sensor should function in both types of ducts.

The need analysis will be concluded at this point, although it is still incomplete. Other relevant issues remain to be identified and studied. For example, we did not describe the requirements on the signal generated by the sensor or the constraints regarding the physical installation in the ducts. Some of the issues mentioned, such as size and compliance with building codes, need to be quantified. The need analysis presented so far is summarized in the following requirements list, which also contains some hypothetical numbers.

6.3 Some Design Requirements

1. Measurement of airflow velocities in the range of 50 to 1000 ft/min.
2. Measurement accuracy: ±30 ft/min.
3. Maximum manufacturing cost: $3 at tens of thousands of units per year.
4. Maintenance-free continuous operation for at least five years.
5. Minimum life: twenty years.
6. Installation in horizontal or vertical ducts.

Many more requirements, covering size, weight, installation, and other aspects, would be listed here if a complete need analysis were performed. Note again that the required manufacturing cost is considerably lower than the target selling price.

6.4 Technology Identification

Looking at existing ways of measuring air velocity and flow rate may provide considerable insight into the physics of the problem and establish various starting points for parameter analysis. Here are several existing devices and methods for measuring flow velocity, presented in no particular order:

Hot-Wire and Thin-Film Anemometers

The basic physics is that of convective heat transfer and electrical-resistive energy dissipation. An electric current through a wire or film placed in the flow is controlled to keep the temperature constant. The faster the flow, the more cooling action is applied to the conductor so that more current is required to produce resistive heating. The current change thus corresponds to the flow velocity. A variation of this implementation is to apply constant current and measure temperature changes.

Two well-known properties of anemometers are their fragility and sensitivity to contamination. The latter weakness is recognized as a potential problem in light of the previous need analysis. Although we could still apply parameter analysis to this physical principle and attempt to overcome the difficulties associated with dust accumulation, we may conclude that such a device would probably turn out to be too expensive and we may decide not to develop this concept further.

Pitot-Static Tube

The basic physics is that of measuring the difference between the static and stagnation (or total) pressures of the flow to deduce its velocity. At first glance, it seems that the need to utilize one or more pressure gages in the design would render it too expensive.

Venturi, Flow Nozzle, and Orifice Meters

Like the Pitot Tube, this is also a Δp device. A constriction accelerates the flow so that potential energy is converted to kinetic energy. The velocity is deduced by measuring the pressure difference between inlet and outlet.

Rotameter

This is a common device used as a flow meter. A float suspended in a vertical tapered tube locates itself at the point where the drag force produced by the upward flow balances the weight. The drag force is proportional to the flow velocity, so the location of the float can be calibrated to indicate this velocity or can directly provide readings of the flow rate.

Turbines and Propellers

The physical principle is utilization of aerodynamic lift and drag forces that the flow exerts on the rotor blades to produce rotational speed, which is proportional to the flow velocity and hence to flow rate. Several methods are available to monitor the rotational speed of the turbine. For example, a permanent magnet and coil can be mounted so that an ac voltage is induced in the coil as the blade passes through the field. The voltage frequency will be proportional to the flow rate.

Particle Marking

This technique is used extensively in wind tunnels to visualize flow. Momentum is transferred from the fluid to particles through drag force. The flow velocity can be measured by timing the motion of a particle between two locations. Two problems associated with applying this technique to the HVAC sensor immediately stand out: it may be difficult and expensive to trace the motion of released particles, and they may contaminate the air fed to the rooms. Furthermore, an appropriate supply of particles is required.

Doppler Effect

An acoustic or optical (laser) device could send a wave of sound or light through the flow and measure the frequency shift in the reflected signal. This shift in frequency correlates with the velocity of the flow. Any device utilizing this principle would undoubtedly cost more than $3 to make.

Drag-Force Flow Meter

This is another aerodynamic drag device. The flow exerts a drag force proportional to the square of the velocity. Measuring the drag force can be used to infer the velocity.

These technologies offer a variety of "initial conditions" for parameter analysis. For conceptual design purposes, the most important aspect of such a listing is understanding the underlying physical principles associated with each technique or device. We now need to scan the list and select the most promising candidates. As often happens in design, this is not a unique process. Different solutions can be derived from different initial conditions. For the present discussion, we will use the last technique, a drag-force device, because its simplicity promises a better chance of satisfying the cost constraint.

6.5 Parameter Analysis

PI: Immerse an object in the flow and measure the drag force exerted on it.

CS: A circular aluminum disk, 2-in. in diameter, is a drag-producing body. This disk is rigidly connected to the top surface of the duct through a flat arm. The drag force can be measured by attaching a strain gage force-measuring transducer, as shown in Fig. 6.1. (Note that the dimensions and material are subject to change later. What is important now is to have a specific configuration to evaluate.)

E: The drag force *F* is given by

$$F = (1/2)\, \rho A C_D\, v^2 \tag{6.1}$$

where ρ is the density of air, *A* is the frontal area, C_D is the drag coefficient, and v is the flow velocity. For a disk normal to the flow, C_D is about 1.1. The density of air is $\rho = 0.075$ lbm/ft³. Neglecting the area of the arm, we calculate the area of the 2-in. disk to be $A = 0.0218$ ft². This gives the following relation between the drag force in lbf and the flow velocity in ft/s:

$$F \cong 2.8 \times 10^{-5}\, v^2$$

For the desired measurement range of 50 to 1000 ft/min, this relation roughly translates into forces in the range 2×10^{-5} to 0.008 lbf. (Note how a 20:1 velocity range translates into a 400:1 range of forces in a drag-force device.)

   To check whether these forces can be measured by the strain gage, we perform a simple analysis of the cantilever beam with the drag force applied at the free end. This shows that the strains range from 0.1 to about 40 $\mu\varepsilon$. These are relatively small strains, so we may consider changes in the design to increase the sensitivity. However, a more serious problem is that strain gages are fragile, sensitive to contamination, and probably too expensive for this design.

PI: Force cannot be measured directly, only through deflection. Strain gages measure very small deflections. A large deflection

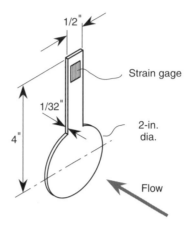

Figure 6.1 The drag force exerted by the flow can be measured by a strain gage as deflection.

may be easier and cheaper to measure. A pendulum can serve as a large-deflection, drag-force device.

CS: A rectangular 6-in. by 1-in., 1/32-in.-gage aluminum plate hinged from the top can constitute the pendulum, as shown in Fig. 6.2. The relationship between the deflection of the pendulum, θ, and the flow velocity, v, should be analyzed, beginning with the drag-force calculation according to Eq. (6.1). The rectangular plate, like the disk, has an approximate drag coefficient of $C_D = 1.1$. The area of the $6'' \times 1''$ plate is $A = 0.0417$ ft². This gives the drag force in lbf as a function of the flow velocity in ft/s:

$$F \cong 5 \times 10^{-5}\, v^2$$

For the desired measurement range of 50 to 1000 ft/min, this relation translates into forces in the range 3.5×10^{-5} to 0.014 lbf.

To find the relation between the drag force and the pendulum deflection angle, θ, we need to analyze the static equilibrium condition of the pendulum, as in Fig. 6.3. Taking moments about the hinge, we find:

$$W\, l \sin\theta = F\, l \cos\theta$$

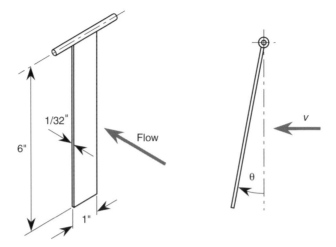

Figure 6.2 A pendulum tilts in response to the drag force exerted by the flow.

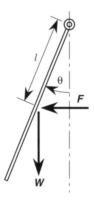

Figure 6.3 Drag, F, and weight, W, are the forces acting on the pendulum in equilibrium.

$$\theta = \tan^{-1}(F/W)$$

The weight of the plate, W, is estimated from the weight density of aluminum (165 lbf/ft^3) and volume ($6'' \times 1'' \times 1/32'' = 0.1875$ in^3 $= 1.09 \times 10^{-4}$ ft^3) at 0.018 lbf. Thus, for flow velocities between 50 and 1000 ft/min and the corresponding drag forces of 3.5×10^{-5} and 0.014 lbf found previously, the inclination angle of the pendulum will be between $\theta_{min} = 0.1°$ and $\theta_{max} = 37°$.

E: As expected from a drag-force device, the range of angles to be measured is quite wide. Especially troublesome is the need to measure 0.1° cheaply, and under the dirty conditions inside the ducts. Note also that the above calculation is approximate because it assumes that the drag coefficient and frontal area do not change with inclination. This degree of accuracy in calculations is sufficient at this stage. However, recognizing that both drag coefficient and area actually decrease at higher angles and flow velocities is important. This understanding leads to identifying the next parameter, or critical issue.

PI: Increase sensitivity at the low end only. Examining Eq. (6.1), which governs the behavior of the pendulum, and in light of the last evaluation step, we realize that the problem of the drag force increasing with the square of the velocity may be circumvented by decreasing other variables in the equation at the same time. The density ρ is constant, but the drag coefficient C_D and frontal area A are sensitive to the angle of the plate, which in turn varies with the flow velocity. Thus, in order to increase sensitivity at the low end only, we need to amplify the effect of the inclination angle by exposing a large area to the flow at low velocities and a much smaller area at high velocities.

CS: A configuration that exposes different areas at low and high velocities may use two pendulums as shown in Fig. 6.4. Note that the downstream plate is split into two so that full exposure to the flow is obtained.

For simplicity of calculations, let us neglect the drag force on the plates when they are almost horizontal. Thus, at the low velocities we make the A_2 area large enough to produce a substantial deflection. For example, using the relations developed earlier, if A_2 is set to about 0.2 ft^2 (two 7-in. × 2-in. plates), the drag force at $v = 50$ ft/min will be about 18×10^{-5} lbf, which is five times more than with the previous rectangular single pendulum. The major improvement, however, is at the high end. Now area A_1 is almost normal to the flow. If we set it to 0.05 ft^2 (a 3.5-in. × 2-in. plate), its drag force at $v = 1000$ ft/min will be about 0.018 lbf, which is only slightly more than with the single rectangular plate. Static equilibrium analysis is now required to

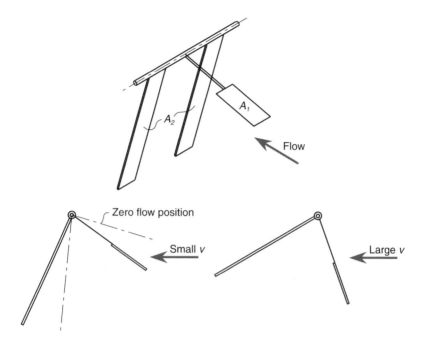

Figure 6.4 The pendulum at the top exposes the large area A_2 at low-flow velocities (lower left) and the small area A_1 at high velocities (lower right).

find the deflections. It will show that the device is somewhat heavy and needs to be made lighter; for example, by reducing the sheet thickness or making them out of plastic. Once an adequate combination of material, plate size, and angle between the plates is found, the design process can proceed with further evaluation.

E: The calculations show that high sensitivity at low velocities and low sensitivity at high velocities can be combined in the proposed configuration. However, the device would function well only when mounted in horizontal ducts. What about vertical ducts?

PI: In the horizontal ducts discussed so far, the pendulum functioned by creating equilibrium between drag and gravitational moments. For use in vertical ducts, a restoring force other than gravity should be used.

CS: The configuration in Fig. 6.5 adds a spring, a counterbalancing mass, and a mechanical stop. At low velocities, the larger A_2 area is almost normal to the flow so the drag force is relatively large,

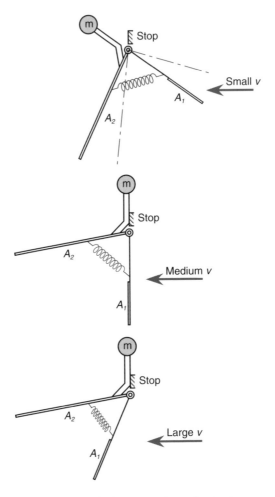

Figure 6.5 A spring, counterbalance, and mechanical stop are added to facilitate operation in vertical as well as horizontal ducts.

producing large deflections. At some medium velocity, the A_2 plates reach the end of their sway, which is defined by the stop. The large drag force that now acts on the smaller plate needs to overcome the relatively high resistance of the spring.

E: With the correct choice of A_1, A_2, m, K (the spring constant), and some additional geometrical characteristics such as arm lengths and initial angular opening, the device can be made to function properly in both horizontal and vertical ducts. The next item still missing is provision for measuring the tilt of the pendulum. In fact, only the deflection of A_1 needs to be measured, relative to a fixed frame of reference.

PI: Measure the absolute angle of A_1 relative to the mounting bracket of the device. This angle sensor should be very inexpensive to keep the overall cost down.

CS: A resistive potentiometer is a very simple, low-cost angle sensor.

E: Potentiometers can easily measure the desired angle; however, low-cost potentiometers usually have high internal friction. A potentiometer with low friction would be too expensive. A potentiometer is therefore not a good choice because it will always have parts that move relative to each other and at the same time need to provide good electrical contact.

PI: In addition to low cost, low friction turns out to be of primary importance with the angle sensor. Only a noncontact device has the potential to cost very little and have no friction.

CS: Use a capacitor-type angle sensor. It is very inexpensive and almost frictionless.

E: We conclude the development of this design now. Several more parameter analysis iterations need to be applied before the conceptual design can be considered complete. For example, we have not yet addressed the problem of minimizing friction in the hinges, the installation of the sensor inside the ducts, or the processing of the angle measurement signals.

6.6 Discussion and Summary

Figure 6.6 summarizes the parameter analysis process used to design the airflow sensor. None of the identified parameters represents "breakthrough understanding" in the problem domain; the designer was only required to utilize fundamental physical principles.

The process described above is not unique. We could start the design process with any of the other initial conditions described in the technology identification section. At any point in the process, different parameters could be identified in the PI step or other configurations suggested in CS. For example, when we decide on a split

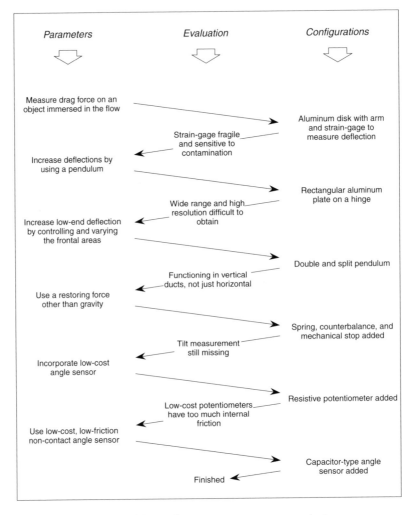

Figure 6.6 Summary of the airflow sensor parameter analysis process.

double-pendulum to increase low-end sensitivity through manipulation of frontal areas, we could identify other parameters. If we decide to increase the sensitivity of the pendulum by regarding it as too stiff a spring (both pendulums and springs deflect more in response to larger forces), we may try to reduce the spring-constant by lengthening the pendulum, reducing its weight, or counterbalancing it with an added mass. This last configuration is sketched in Fig. 6.7. The mass creates a moment about the hinge that joins the

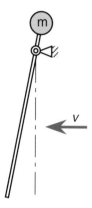

Figure 6.7 Counterbalancing a pendulum increases its sensitivity as a flow-velocity sensor.

drag force in producing a relatively large deflection, thus increasing the overall sensitivity. Note, however, that sensitivity at the high end also increases with this arrangement.

An alternative parameter would be to decrease the sensitivity at the high velocity while not changing sensitivity at low speeds. Still having a spring model for the pendulum in our minds, we realize that a spring-like device may be made stiffer by adding another spring to it in parallel, as in Fig. 6.8.

Here, a torsion spring was added to produce the following effect: When there is no flow and the pendulum is vertical, the spring is unstressed. When the flow is low, the influence of the spring is hardly noticeable because small deflections produce only small spring forces. When the flow is high, however, the increased deflection magnifies the resistance of the spring, resulting in smaller pendulum deflections at the high end of the velocity range. Now that the overall range of deflections has shrunk, we may want to continue the parameter analysis process to increase overall sensitivity so that the small deflections would not be too difficult to sense.

Another conceptual design path may emerge from the "low-cost, low-friction angle measurement" parameter. In the above discussion, a capacitor-based sensor is selected. But the same parameter could result in a multitude of other configurations such as an optical encoder (one or two pairs of LED-type emitters and

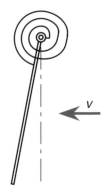

Figure 6.8 Adding a torsion spring reduces the deflection at high velocities more than at low velocities.

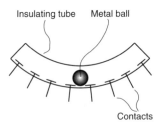

Figure 6.9 A proposed design for an angle sensor. The metal ball short-circuits different pairs of contacts depending on the tilt angle.

receivers placed on opposite sides of a slotted disk), or a magnetic sensor in which geometrical changes vary the intensity of a magnetic field. Another possibility is to design our own low-cost, low-friction angle sensor similar to the one shown in Fig. 6.9, where a metal ball short-circuits pairs of electrical contacts:

6.7 Thought Questions

1. Modern computer-controlled automotive engines employ an air-flow sensor to match the quantity of fuel injected with the

amount of air. Investigate how these sensors work and whether the same operation principle can be used for HVAC applications.

2. One flow-velocity sensing device mentioned in the technology identification section (6.4) is the anemometer. However, we ruled it out due to sensitivity to contamination. Apply parameter analysis to the design of a low-cost anemometer, with emphasis on eliminating this sensitivity.

3. Another method mentioned under technology identification is the use of a rotameter. The design of these devices for upward flow measurement is straightforward. Can you design a similar device for horizontal and downward flows?

4. In the discussion and summary section (6.6), we mentioned several alternative solutions to those adopted in the main design process (Section 6.5). Repeat this process using the alternative parameters and configurations to generate more airflow sensor designs based on the basic pendulum and drag-force concept.

5. As a way to practice parameter analysis, choose any of the methods listed under technology identification as a starting point for a new airflow sensor design process.

6. Locate available low-cost angle sensors—such as potentiometers, capacitance- and inductance-based, magnetic, and encoders—in commercial catalogs and compare them based on accuracy, durability, and cost.

6.8 Bibliography

A source of information on HVAC systems is required for a good need analysis and to formulate the requirements. For example:

ASHRAE Handbook—Fundamentals. Atlanta: The American Society of Heating, Refrigerating and Air-Conditioning Engineers, 1997.

McQuiston, F. C., Spitler, J. D., and Parker, L. *Heating, Ventilating, and Air Conditioning: Analysis and Design.* 5th ed. New York: John Wiley & Sons, 2000.

Many books on measuring techniques are available and will be useful to a designer working on sensor design. One of the most comprehensive is:

Doebelin, E. O. *Measurement Systems: Application and Design.* 4th ed. New York: McGraw-Hill, 1990.

A general handbook with good in-depth coverage of many topics is:

Avallone, E. A. and Baumeister, T. (eds.), *Marks' Standard Handbook for Mechanical Engineers.* 10th ed. New York: McGraw-Hill, 1996.

Automotive-specific information that can help with thought question no. 1 may be found in:

Bosch, Robert. *Automotive Handbook.* 4th ed. Cambridge, MA: Bentley Publishers, 1997.

and in books published by the Society of Automotive Engineers (SAE).

Information about commercial sensors is widely available on the Internet, in publications such as *Sensors* magazine, and in catalogs of companies such as Edmund Scientific Co. and Omega Engineering, Inc.

7

Conceptual Design Case Study: Cut-Edge Sensor for Flooring Removal

This chapter presents another case study of sensor design. The sensor guides an asbestos-flooring removal machine along the cut edge of the flooring material once the first strip has been removed. As with many sensor design tasks, a multitude of technologies can be used. Noncontact imaging is selected for this example because of the harsh operating environment and the need to avoid accidental release of asbestos fibers into the air.

7.1 The Initial Need

An asbestos-flooring removal machine is shown in Fig. 7.1, whereas a cross-section through a typical floor is presented in Fig. 7.2. The machine was designed to assist in removing old flooring material without releasing harmful asbestos fibers into the air. It is equipped with heating elements that radiate enough heat (about 1500 W) to raise the temperature of the adhesive to around 100°C. Then, the operator makes a cut in the flooring and removes a 30-cm wide strip by pulling on it. Some of the molten adhesive remains on the subfloor, but some is still attached to the removed layer, thus containing the hazardous asbestos fibers and preventing them from being dispersed in the air. The process of heating the adhesive layer is relatively slow: the machine traverses the room at a speed of 3 m/hr. To save operator time, it is desirable to design a sensor that will allow

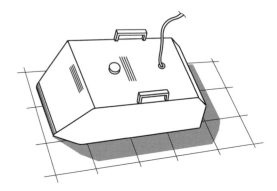

Figure 7.1 The existing flooring removal machine.

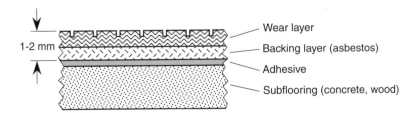

Figure 7.2 A cross-section in typical asbestos-containing flooring.

the machine to steer itself along the cut edge of the flooring once a first strip has been removed.

7.2 Abbreviated Need Analysis

Performance

As with any sensor design, the quantity to be measured needs to be quantified first. The cut edge to be followed results from a person making a cut with a utility knife at an approximate distance of 30 cm from a wall or a previously cut edge. The cut edge may therefore be somewhat skewed, jagged, or wavy, as shown in Fig. 7.3. Reasonable assumptions to quantify the cut edge are a maximum waviness of 3 cm (amplitude of 1.5 cm) and a maximum jaggedness of 0.5 cm. The height of the cut edge is 1 to 2 mm above the subfloor, but it may be tilted sideways; that is, it may form angles other than 90°

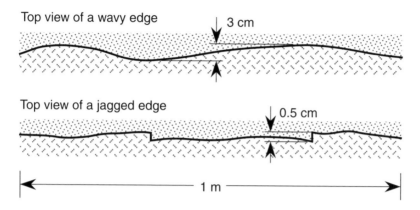

Figure 7.3 Characterization of wavy and jagged cut edges in the flooring.

with the floor when viewed from the front. We set this deviation to a maximum of ±30° from the vertical.

When use of the sensor-guided machine is considered, it is revealed that too accurate a sensor will cause the machine to closely follow the irregularities in the cut edge, which is undesirable. What is needed is some averaging effect where local variations and deviations are ignored. Achieving this by mathematical averaging of several sensor readings is clearly overkill. The same result can be obtained by designing a low-sensitivity sensor in the first place. Thus, we set the desired resolution to 1 cm, meaning that deviations smaller than 1 cm to either side of the machine heading should not produce a change in the sensor reading. If the sensor functions well, the deviation from the nominal course will never exceed about 1 cm. However, to be on the safe side, we may want the desired range of measurement to be ±3 cm from the center.

Some aspects of the operation of the machine are relevant to the functioning of the sensor. The low speed of 3 m/hr implies that the sensor can have a slow response time and still provide adequate positioning accuracy. Consider, for example, the unlikely case of the machine deviating by a full 90° from its course. If the sensor takes 10 seconds to recognize it, the deviation will still be less than 1 cm, or 3% of the cut-strip width.

The high temperatures involved in the flooring removal operation may pose a problem to the sensor. When the adhesive reaches

100°C, the top surface of the wear layer may be at 220°C, and the sensor should be able to endure this condition. In addition, the surfaces over which the machine operates may be characterized as follows: the subflooring is wood or concrete of variable color, covered with soft and hard adhesive residue. The wear layer of the yet-to-be-removed flooring may have any color and pattern printed or embossed on it. Such patterns may also consist of grooves as deep as one-half the total thickness, that is, 0.5 to 1 mm deep.

As does the machine itself, the sensor should also use regular electrical household current for its operation. In order not to subtract from the power available for the heating task, the power consumption of the sensor will be limited to 100 W.

Value

Freeing the human operator from the task of continuously monitoring and guiding the flooring removal machine is the major benefit of the cut-edge sensor. With a sensor-equipped machine, the operator will be able to attend to other tasks or place several machines in different locations to operate simultaneously. The monetary value of the sensor can therefore be established by considering the increased productivity. For the present discussion, we recognize that the machine itself is a relatively expensive device with a retail cost of around $2000, so the target production cost for the sensor is set at $100.

Size

Since the bottom of the machine is covered with heating elements, possible sensor locations, which offer downward accessibility, include the front, back, and sides. Installation in the front is probably preferable in terms of the timely detection of deviations in course. There does not seem to be a reason to severely limit the physical size or weight of the sensor. It should not be as bulky or heavy as to hinder the performance of the machine, however, so we set its maximum size to a 20 × 20 × 20-cm cube, and its maximum weight to 30 N (3 kg). We also note that the sensor may be stored and transported separately from the machine, and mounted and plugged in on site.

Safety

Besides provisions normally expected from a device used near people, the most important safety requirement is to comply with the overall goal of the machine; that is, the sensor should not contribute to releasing any asbestos fibers from the floor. This requirement is, of course, of utmost importance.

7.3 Some Design Requirements

1. Measurement of the location of the 1–2-mm high cut edge with respect to the machine heading. The characteristics of the cut-edge geometry, subflooring, and removed layers are as described in the need analysis.
2. Measurement resolution: no smaller than 1 cm.
3. Maximum deviation to detect: 3 cm to either side of the cut edge.
4. Toleration of surrounding air temperature of 100°C.
5. Maximum power consumption at household electricity: 100 W.
6. Maximum manufacturing cost: $100 at an expected production volume of 500 units per year over a 10-year period.
7. Maximum size: $20 \times 20 \times 20$ cm.
8. Maximum weight: 30 N.
9. Not causing the release of any asbestos fibers during operation.

7.4 Technology Identification

We may divide various position-sensing techniques into two broad categories: contact and noncontact. A mechanical follower idea that can be used as a cut-edge sensor is shown in Fig. 7.4. The problem with using contact sensors in this application is that they need to follow a sticky edge covered with hot adhesive and may also release asbestos fibers from the backing layer. A much more promising venue to explore is noncontact sensors.

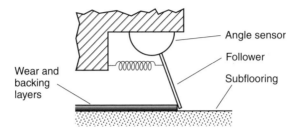

Figure 7.4 A contact sensor based on a mechanical follower.

Noncontact sensing can be based on various principles. We may detect the cut edge by performing two range measurements, one to the wear layer and one to the subfloor. The difference in range is 1–2 mm, so an accurate range finder will be required. We may look into range measurement by timing the flight of sound waves, as may be done in an auto-focus camera, for example. The required sensor resolution (1 mm corresponds to about 3.6×10^{-7} second) is much higher than that in cameras, however, and may be expensive to implement.

A different noncontact technique is imaging, where the intensity of light reflected from the surfaces is measured. It is unclear at this stage whether a distinct image of the cut edge can be created. Nevertheless, this principle seems more promising than range sensing, and it is therefore selected as a starting point for parameter analysis.

7.5 Parameter Analysis

PI: Measure the different light intensities bouncing back from the wear layer and subflooring.

CS: The configuration in Fig. 7.5 uses a light bulb and two one-dimensional arrays (rows) of photosensitive elements. The bulb is shielded so that its light does not reach the optical detectors directly. Each array senses a different light intensity because the light reflects from different surfaces. if the machine deviates from its intended course, the variation in light intensity will produce an error signal.

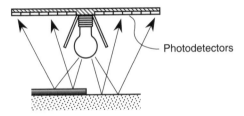

Figure 7.5 A configuration to sense the intensity of light reflected from different surfaces.

E: The reflected light may not vary enough to determine the location of the edge because of the presence of adhesive residue and dirt and the possibility of different material and color combinations for the surfaces.

PI: Create a greater difference in light intensities. If the images of the wear layer and subflooring are indistinguishable from each other, create and sense a third image—the shadow of the cut edge.

CS: Position the bulb away from the cut edge and let its light shine at an angle across the edge so that a shadow is cast, as shown in Fig. 7.6. The shadow will be detected as a dark line on a single optical detector array.

E: Now that the image of the shadow is the quantity to be sensed, we should try to make it large enough for easy detection. Also, the cut edge may be inclined at up to 30°, so the angle of incidence of the light (the angle between an incident ray and the normal to the surface) should be greater than 30° to produce a shadow.

PI: The size of the shadow depends on the angle of incidence. Increasing this angle will make the shadow easy to detect.

CS: The configuration of Fig. 7.7 moves the bulb further away from the cut edge to cast a large shadow. The detector array is moved and tilted accordingly. At the nominal position, with the image of the shadow at the center of the detector, the angle of incidence of a light ray hitting the cut edge is about 60°.

E: The shadow will not be clearly seen on the photodetector because the light bouncing from the floor will diffuse and scatter in all directions.

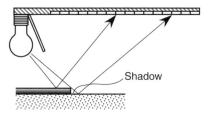

Figure 7.6 Lighting the cut edge at an angle creates a detectable shadow.

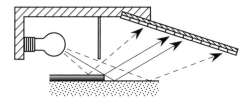

Figure 7.7 Increasing the angle of incidence of the light enlarges the image of the shadow.

PI: Focus the reflected light by using a lens.

CS: A lens is added in front of the detector to focus the image of the cast shadow. For the required performance, a low-cost plastic lens, as used on single-use cameras, may suffice. The configuration in Fig. 7.8 integrates the lens with the detector in one housing. A separate shield between the bulb and detector is still required to prevent direct illumination of the detector. The focusing lens (convergent, or convex, lens) allows the use of a smaller detector array.

At this point we need to calculate some dimensions to match the performance of the design with the required resolution and range of detection. Figure 7.9 shows the information pertinent to using a convergent lens.

The fact that the image is inverted relative to the object should not pose a problem because this can easily be accounted for in the signal-interpretation step. The magnification of the lens, that is, the ratio of image size to object size, is given by b/g. Another important relationship is the following approximation:

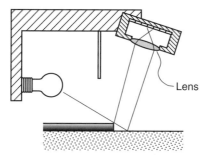

Figure 7.8 Adding a lens in front of the photodetector to create a sharp image of the shadow.

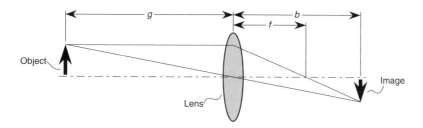

Figure 7.9 Basic geometrical properties of a convergent lens (g = object distance, b = image distance, f = focal length).

$$\frac{1}{f} = \frac{1}{g} + \frac{1}{b}$$

There are two issues to consider when determining dimensions for our design: the spacing between the sensing elements and the thickness of the image of the shadow on the detector. If the actual detection range (±3 cm = 60 mm) is converted to an imaging range of 20 mm, the magnification is set at one-third. From the constraints on the size of the sensor, we decide to choose $g = 75$ mm and $b = 25$ mm, giving the desired magnification. The focal length of the lens is thus determined to be 18.75 mm.

The height of the cut edge varies between 1 and 2 mm, and the shadow is cast by shining light at 30° to the horizon. The "thickness" of the object to be sensed is therefore twice the height, or 2 to 4 mm. With the one-third magnification, this

131

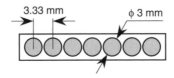

Figure 7.10 The seven-photodiode linear array used as detector.

translates into an image 0.67- to 1.33-mm wide. In order to keep the resolution low enough, that is, not detect deviations smaller than 1 cm, we should use six sensors in the array. For reasons of symmetry, however, seven sensing elements are a better choice. We also decide now to use 3-mm diameter photodiodes as the sensing elements. The layout of the detector is shown in Fig. 7.10.

E: The signal-to-noise ratio may be too small due to the presence of ambient light. The signal is the image of the shadow, created by different reflected light intensities. Light from other sources in the room, especially the heating elements on the machine, constitutes the noise.

PI: A larger signal-to-noise ratio can be obtained either by decreasing the noise or increasing the signal. To decrease the noise, additional physical shielding could be used to further enclose the bulb, cut edge, and detector. A simpler solution, however, may be to increase the signal by using a light source whose wavelength is different from that of the noise.

CS: The heating elements emit light whose infrared content is relatively high. Operating the bulb–detector pair at the other end of the spectrum—the ultraviolet (UV) wavelength—would help separate the signal from the noise. Specifying a UV bulb and detector, or a lens-mounted filter that blocks all but the UV waves, is straightforward.

E: The configuration developed so far seems viable; however, what would guarantee detection of the cut-edge shadow and not shadows produced by grooves in the wear layer?

PI: The depth of the wear-layer grooves is about one-half the height of the cut edge. We may try to use this difference to distinguish between the two, but adhesive residue on the sub-

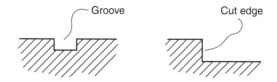

Figure 7.11 The difference between a groove (left) and the cut edge (right).

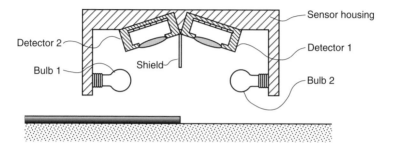

Figure 7.12 A design that comprises two multiplexed bulb–detector pairs.

flooring may cause the cut edge to appear shallower than it really is. A better concept is based on the realization that grooves are symmetrical while the cut edge is one-sided, as shown in Fig. 7.11. This property can be used to cast shadows by shedding light from different directions.

CS: If the design is doubled and its operation made to alternate (multiplex) between the bulb–detector pairs, then the groove will always produce a shadow while the cut edge will be detected on alternate readings only. A configuration with this arrangement is shown in Fig. 7.12. A groove on the wear layer would produce a signal (shadow) on both detectors, while the cut edge would show on detector 1 only.

E: Most major issues have been dealt with and the solution seems feasible, so we conclude the conceptual design development at this point. The sensor will be mounted on one of two brackets provided on the front of the flooring-removal machine. One bracket will align the center of the sensor with the left side panel of the machine and the other, with the right side panel.

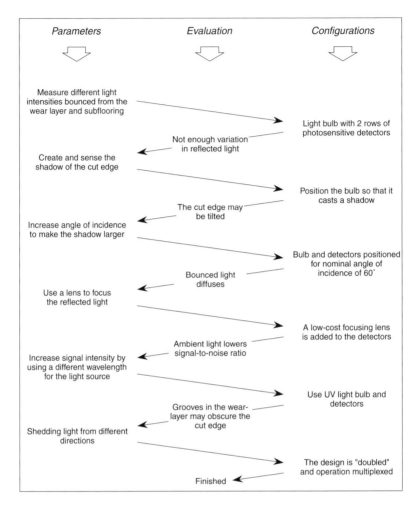

Figure 7.13 Summary of the cut-edge sensor parameter analysis process.

7.6 Discussion and Summary

The parameter analysis process of designing the sensor is summarized in Fig. 7.13. It shows how the simple concept of measuring the intensity of light bounced from the floor has evolved into a device that actually creates a new image—a shadow of the cut edge—and senses it. The quantity to be measured by the sensor can thus be enhanced and distinguished from the surrounding noise, which is in the form of grooves and other surface features and ambient light.

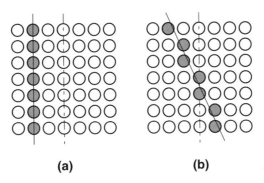

(a)   **(b)**

Figure 7.14 Images of the cut-edge shadow on a two-dimensional photodetector array: (a) parallel course deviation, (b) angular deviation.

Several secondary issues regarding the design should still be addressed. A decision needs to be made whether to use one-dimensional detector arrays (a single row of photodiodes) or two-dimensional grids. The signals from the detectors should be conditioned (amplified, filtered, and possibly digitized) and fed to a controller for the steering motor. A control strategy needs also to be developed, specifying how error signals will be converted to course-correction instructions. For example, Fig. 7.14 shows two possible signals recorded on a two-dimensional detector array. The one in (a) represents a parallel deviation, while the signal in (b) shows an angular deviation. It may also be desirable to calculate the velocity of deviating from the nominal course by taking two successive position readings over a given time period.

7.7 Thought Questions

1. Examine noncontact range-sensing technologies and compare them in terms of their ability to provide a 1- to 2-mm measurement resolution at a reasonable cost.

2. Design a noncontact sensor for the application of this chapter based on a range-measuring principle.

3. In the technology identification section (7.4), we mentioned that contact sensors had two major problems: they would be touching

a sticky and hot cut edge, and they might cause the release of asbestos fibers. Design a contact sensor for this application that would overcome these two difficulties.

7.8 Bibliography

As in the previous chapter, general books on measuring techniques will come in handy while designing a sensor. The following book is again suggested:

> Doebelin, E. O. *Measurement Systems: Application and Design.* 4th ed. New York: McGraw-Hill, 1990.

When the choice was made to use imaging, books on optics proved useful. Many such books are available, so none in particular will be mentioned here. A manufacturer or retailer catalog, such as the one below, should be consulted for availability of low-cost lenses and photodetectors:

> Edmund Scientific Co. *Edmund Industrial Optics.*

8

Conceptual Design Case Study: Low-Cost Industrial Indexing Systems

This chapter describes low cost and flexibility as two key requirements in the design of a new drive system for a mechanical indexer. Need analysis helps to generate two key insights into the design task. Based on these two insights, two competing designs are developed using the parameter analysis methodology to address the requirements. A detailed discussion of technical issues involved in realizing these concepts is presented. Both of the designs were built and tested for performance.

8.1 The Initial Need

Indexing systems are often used in industry to incrementally move products along an automated line and stop them at predetermined workstations. At these workstations, people or machines perform assigned tasks, such as welding or assembly of components. Two common technologies for producing incremental motion are servo drives and mechanical indexers. Servo drives employ a dc motor, speed reducer, clutch–brake system, sensors, and electrical controls. They are highly flexible in the sense that they can accommodate changes in the product line by adjusting the timing sequence for various activities. In contrast, mechanical indexers constitute rigid automation. They use a continuously running electric motor, speed reducer, clutch–brake system, and indexer unit. The indexer unit converts the constant input motion of the electric motor into intermittent output motion (linear or rotary) by using a cam–follower mechanism. Both of these indexing technologies are expensive.

Market research has indicated that there is a need for flexible, yet moderately priced, indexing systems. The mechanical indexer technology would capture a large segment of the low-end market if it could satisfy cost and flexibility requirements. In addition, typical applications in this market segment are usually not very sensitive to the acceleration profile of the load. The goal of this design task is therefore to develop a low-cost drive system for mechanical indexers.

8.2 Abbreviated Need Analysis

The black-box methodology with inputs and outputs (see Section 2.2) can be used to precisely identify the need. In this particular case, the input is ill-defined because several types of energy inputs are possible. The need is thus formulated by looking at the output of the black box and requiring that it provide energy to the shaft driving the mechanical indexer. Several questions come to mind in this context:

1. What is the energy requirement for a typical mechanical indexer? Is it constant, or does it vary with time?
2. How much energy or power is required?
3. What are the characteristics of an ideal solution?
4. What are the possible energy sources?
5. Where is the cost locked up? Or what are the primary cost drivers in the current design?

Performance

The primary function of the drive system is "to supply energy" to the indexer. Because the indexer continuously cycles through acceleration, deceleration, and dwell (stop) periods, the required power and torque do not stay constant. To determine them, an ideal acceleration profile for the output load must be determined. This profile can then be used to find the velocity, displacement, and torque profiles.

Several acceleration profiles (constant, trapezoidal, modified trapezoidal, and modified sinusoidal) can be used to accelerate and decelerate the load. Most indexers use the modified sinusoidal acceleration profile since it reduces jerk, peak acceleration, and peak velocity. By decreasing the peak velocity, this profile reduces the amount of kinetic energy that has to be dissipated each time the assembly line is brought to a stop. Note that although the final solution may not produce the modified sine acceleration profile, this assumption provides an initial starting point for determining torque requirements and, later, sizing parts. It also helps in the evaluation of competing conceptual solutions when they satisfy all other requirements to the same extent.

Based on the equations of a modified sinusoidal acceleration, a plot of the angular acceleration and velocity of the load for different positions of the input shaft is shown in Figs. 8.1 and 8.2. It is assumed in these calculations that the speed of the input shaft is constant.

The output torque of the drive system is the product of the angular acceleration (α) and inertia (*I*) of the load. However, to design the drive system, the input torque required by the indexer must be characterized. The indexer can be regarded as a speed reducer with variable gear ratio, where the instantaneous gear ratio is the output shaft speed divided by the input shaft speed. Based on this insight, the input torque to the indexer can be obtained by multiplying the output torque of the drive system (*Iα*) by the gear ratio calculated from Figs. 8.1 and 8.2. The resulting input torque profile for the indexer is shown in Fig. 8.3. An ideal drive system must closely emulate this torque profile.

Let us now estimate the energy requirement of the indexer. At an abstract level, the indexer system accelerates an object from a state of rest to the desired velocity and then decelerates it back to rest. Neglecting frictional losses, we find that the net energy requirement is therefore zero. Close examination of the input torque profile (Fig. 8.3) reveals this fact: The area under the curve during the acceleration period is numerically equal to the area during the deceleration period, and these areas represent energy. Energy supply

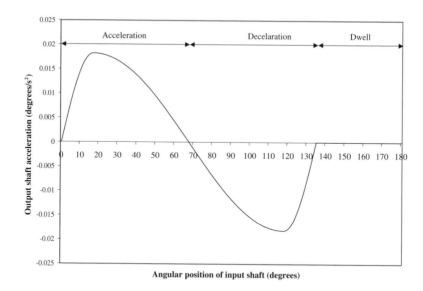

Figure 8.1 The modified sinusoidal acceleration profile of the output shaft.

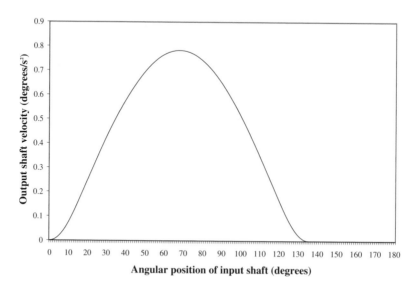

Figure 8.2 The output shaft velocity profile obtained by integrating the acceleration.

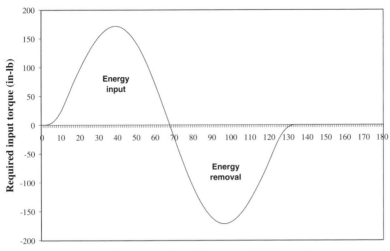

Figure 8.3 The required input torque for the indexer.

is needed to drive the load in the acceleration phase, but this energy must be removed from the system during deceleration.

Value

Cost analysis reveals that over 50% of the product cost comes from the electric motor and the speed reducer. For the new drive system to compete on the basis of cost effectiveness, it must eliminate these components. Power consumption is relatively insignificant for these drive systems. To reduce the required initial investment by the customers, the manufacturing cost should not exceed $400.

In this example, we are not generating a range of performance versus value estimates since the target performance of the device is somewhat narrowly defined (essentially reducing cost). In a more complete need analysis, we would estimate the value to the user as it varies over a range of performance specifications, depending on the intended application of the indexer.

Size

Existing indexers have proved to be extremely reliable, with a long life. Many units, however, may require replacement of their drive system. It is therefore decided that the new drive system must interface with the existing indexers. This constraint imposes geometric requirements, such as the height of the drive shaft from the base and the maximum available space.

8.3 Some Design Requirements

1. Power an indexer with a peak torque of 180 in-lb.
2. Provide the ability to adjust the timing sequence for various activities.
3. Maximum manufacturing cost: $400 at low production volume.

These are three key requirements. Additional requirements concerning other aspects, such as size, should be included in the complete list of requirements.

8.4 Technology Identification

Two ideas, based on the insights gained in the need analysis phase, are:

1. *Pneumatic actuation.* Electric motors typically run at high speeds, thus requiring a step-down speed reducer. Linear actuators, particularly pneumatic cylinders, may be a low-cost alternative as a power source. Since they are widely used in industry, air supply is readily available. They can be actuated on-demand and are also relatively slow. Power from a pneumatic actuator may be used to continuously rotate the drive shaft of the indexer.
2. *Spring energy storage.* The net energy supply required to drive the indexer is very small—it equals only the frictional losses. A smart approach would be to store the dissipated energy during the deceleration phase and release it for accelerating the object. The actuator will then need to supply only a small amount of energy to make up for the frictional losses. A spring may be used for such energy storage.

These two ideas serve as initial conditions for the two independent parameter analysis processes described in the next two sections.

8.5 Parameter Analysis of Conceptual Design I

PI: The first conceptual design process starts with the pneumatic actuator idea. For the actuator to drive the shaft, its linear motion should be converted to the rotary motion of the shaft.

CS: The configuration in Fig. 8.4 uses a pneumatic actuator, cable, and pulley mounted on the drive shaft to convert motion from linear to rotary. During the actuation stroke, the piston moves from right to left while pulling the cable. The cable in turn rotates the pulley and the drive shaft in the counterclockwise direction. Note that at this stage in parameter analysis, the configuration is not fully functional. This is an acceptable practice because the goal is to have an initial configuration that can be evaluated and improved.

E: During the return stroke the piston tries to push the cable. Therefore, the cable does not return to its initial position.

PI: Tension must be maintained in the cable during the return stroke.

CS: A tension spring can return the cable to its original position. The new configuration incorporating the spring is shown in Fig. 8.5. The cable moves from right to left during the actuation stroke while rotating the shaft counterclockwise and extending the spring. The tension spring pulls the cable back to its original position during the return stroke.

E: This configuration produces an oscillatory motion of the shaft. During the actuation stroke (the piston moves left), the drive shaft rotates counterclockwise, but it reverses direction during the return stroke (indicated by dashed arrows). This is obviously undesirable because the drive shaft needs to continuously rotate in one direction.

PI: Eliminate the clockwise rotation of the shaft during the return stroke of the actuator.

CS: A one-way clutch can be used to transmit torque in one direction while allowing free running in the opposite direction. Two

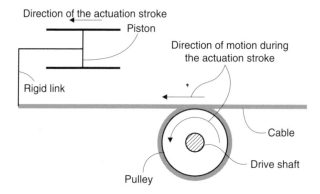

Figure 8.4 The linear motion of the actuator is converted to shaft rotation by a cable and pulley.

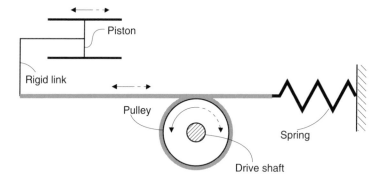

Figure 8.5 The spring restores the cable to its original position during the return stroke of the actuator.

popular embodiments of a one-way clutch are the sprag and roller-ramp types shown in Fig. 8.6. When the driving outer race rotates in the locking direction, sprags or rollers wedge between the inner and outer races, thus transmitting the torque. When the direction of rotation is reversed, the sprags or rollers disengage and allow free relative motion between the races.

The configuration is now modified to incorporate a one-way clutch between the pulley and the drive shaft, as shown in Fig. 8.7. The clutch will transmit motion and torque between

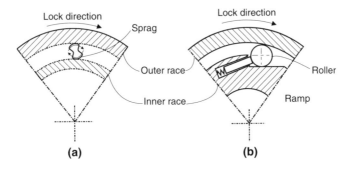

Figure 8.6 (a) A sprag one-way clutch and (b) roller-ramp one-way clutch.

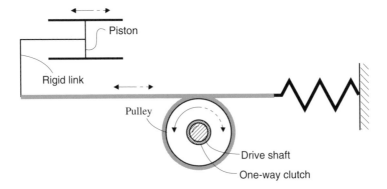

Figure 8.7 A one-way clutch is added to eliminate clockwise rotation of the drive shaft.

the pulley and the shaft in the counterclockwise direction and free wheel in the clockwise direction. This will ensure unidirectional rotation of the drive shaft.

E: The new configuration does not utilize the return stroke of the actuator. This creates unnecessary idle time and limits the overall speed.

PI: Use the return stroke for powering the drive shaft continuously.

CS: A system of two one-way clutches can be employed as shown in Fig. 8.8. When the pneumatic actuator moves from right to left (marked by solid arrows), pulley #1 rotates in the counterclockwise direction and transmits the torque to the drive shaft. At the same time, pulley #2 free wheels as it rotates clockwise.

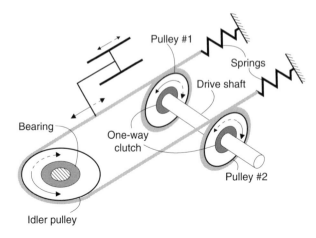

Figure 8.8 Two pulleys in parallel continuously power the drive shaft.

During the reverse stroke of the actuator (dashed arrows), pulley #2 transmits the torque while pulley #1 rotates freely. Two
springs are used to maintain tension in the cable.

E: The springs add complexity to the design as additional parts.
They also introduce continuously changing forces, which
depend on the instantaneous deflection of the springs. These
variable forces subtract from each other to cause continuously
decreasing torque on the drive shaft. The resulting torque profile is not the desired one.

PI: The cable moves back and forth in accordance with the movement of the actuator, and the springs always deflect opposite
each other (one shortens as the other elongates). A fixed-
length, closed-loop cable would do exactly the same while eliminating the springs.

CS: A configuration that employs a closed-loop cable is shown in
Fig. 8.9. The operation of this design is identical to that of the
previous configuration except for the tension springs.

E: This design eliminates the need for springs, but it violates size
constraints. It is also more expensive because it requires an
additional idler pulley and associated shaft and bearings.

PI: Improve overall compactness and reduce the number of components.

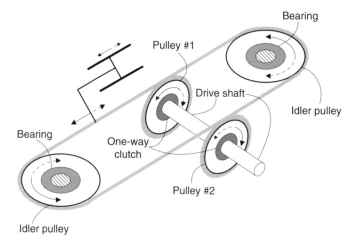

Figure 8.9 The springs are replaced by a closed-loop cable running between two idler pulleys.

CS: A single shaft can be used to mount both idler pulleys by repositioning them, as shown in Fig. 8.10. If the design is made compact and uses conventional chains or belts as cable, considerable twisting of the latter may produce undesirable effects and short life. We therefore decide to use a 3-D chain (see detail in Fig. 8.10), which is very versatile and can be arranged between pulleys at different angles. The chain teeth positively engage the sprockets, thereby eliminating any possibility of slip. Furthermore, the overall size of the design is now greatly reduced.

E: This configuration succeeds in replacing the costly electric motor and the speed reducer. However, it still requires a brake to bring the load to rest. A commercially available brake will need to be added. Otherwise, this conceptual design process, summarized in Fig. 8.11, is complete.

8.6 Parameter Analysis of Conceptual Design II

PI: The second conceptual design is based on the spring energy storage idea recognized in the technology identification step: Use a spring to accelerate and decelerate the drive shaft.

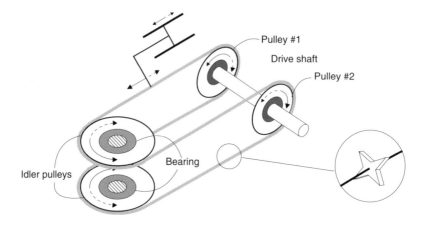

Figure 8.10 Rearranging the idler pulleys eliminates one shaft. A 3-D chain (see detail) is specified.

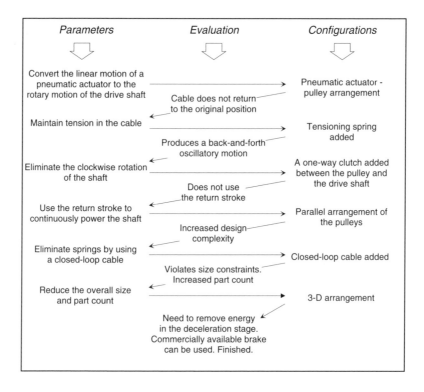

Figure 8.11 Summary of the parameter analysis process for conceptual design I.

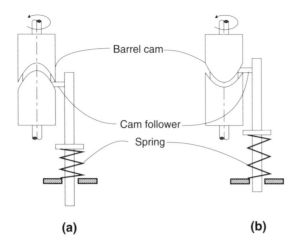

Figure 8.12 A barrel cam mechanism shown in (a) the fully compressed and (b) extended positions of the spring.

CS: A barrel cam can convert the force of a helical compression spring into rotary motion of the drive shaft, as shown in Fig. 8.12. The spring is initially compressed (Fig. 8.12a) and therefore stores potential energy. This stored energy is released during the acceleration phase of the modified sine profile (Fig. 8.1) until the spring reaches its fully extended position (Fig. 8.12b). Next, the rotation of the drive shaft compresses the spring, thereby transferring the energy back to it, and the cycle repeats itself.

E: Barrel cams are difficult to manufacture and may be too expensive.

PI: Plate cams are cheaper to manufacture than barrel cams, so use a plate cam for the motion conversion.

CS: Figure 8.13 shows a plate cam mechanism. The operation is identical to that of the barrel cam mechanism. The cam can be initially sized by performing an energy balance (neglecting frictional losses):

$$F_{spring} \cdot dr = T \cdot d\theta$$

where F_{spring} is the force exerted by the spring (a function of its displacement), dr is the incremental change in radius, T is the

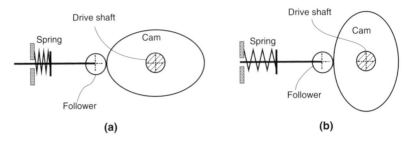

Figure 8.13 A plate cam configuration shown in (a) the fully compressed and (b) extended positions of the spring.

time-dependent torque required to drive the indexer, and $d\theta$ is the incremental change in the angle of the drive shaft. Assuming initial values for the spring rate, initial compression, and base radius of the cam, the cam profile can be determined. The preliminary calculations show that a small elliptical cam of about 4″ major diameter and a spring with a 75 lb/in. rate can drive the indexer.

E: The sizing calculations prove the viability of the design. However, frictional losses were neglected. Because there is no external energy input into the system, the friction forces will prevent the cam from reaching its original position; that is, the fully compressed position of the spring.

PI: Add an external energy source to make up for frictional losses.

CS: A pneumatic actuator would be a cost-effective energy source. A modified configuration incorporating a pneumatic actuator is shown in Fig. 8.14. The operation of this design consists of three distinct phases. The initially compressed spring (Fig. 8.14a) drives the cam attached to the drive shaft through 90°, reaching its fully extended position (Fig. 8.14b). The cam profile can be determined by treating an infinitesimally small section of it as an inclined plane and resolving the spring and frictional forces. Since there are no impact loads, the profile can be made to match the modified sinusoidal acceleration. Next, the torque from the inertial load compresses the spring by rotating the shaft through another 30° (Fig. 8.14c). The cam profile can be determined as before to match the modified sinusoidal acceleration profile. Finally, the pneumatic

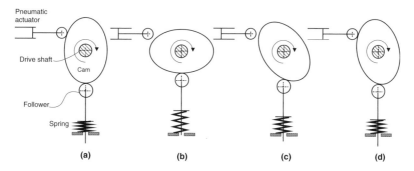

Figure 8.14 The spring is in the fully compressed position (a) when it starts rotating the cam and reaches its fully extended position (b). The load inertia compresses the spring further during the deceleration period (c), when the pneumatic actuator kicks in (d).

actuator rotates the shaft by 60° (Fig. 8.14d), bringing the spring back to its fully compressed position. A simple mechanical switch triggered by the rotation of the drive shaft times the actuator.

E: The spring supplies a significant portion of the energy during the acceleration phase, so it is important to stop the drive system in the fully compressed position of the spring. Otherwise, the actuator will not be able to provide the required energy to accelerate the system. When the spring starts from rest in the fully compressed position, the direction of cam rotation is unpredictable.

PI: Stop the cam slightly after the fully compressed position of the spring. This will ensure rotation in the correct direction.

CS: A solenoid-released latching mechanism can be used to stop the follower arm, as shown in Fig. 8.15. The latch is spring-loaded and therefore locks the follower when the groove in the follower arm aligns with the latch. When the solenoid is actuated, it pulls the latch back, thus releasing the spring.

E: Locking the follower will not stop the rotation of the drive shaft since the inertial load can still rotate the cam, as shown in Fig. 8.16.

PI: To stop the system, lock the cam or drive shaft directly.

151

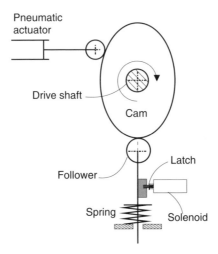

Figure 8.15 The latch mechanism in the locked position.

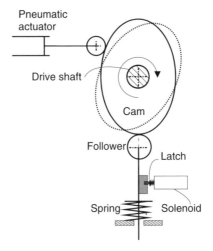

Figure 8.16 Drive shaft rotation due to the inertial load is possible even when the follower is locked.

CS: The latch mechanism is positioned to lock the drive shaft directly.

E: All critical parameters have been addressed, so the parameter analysis process is concluded and summarized in Fig. 8.17.

152

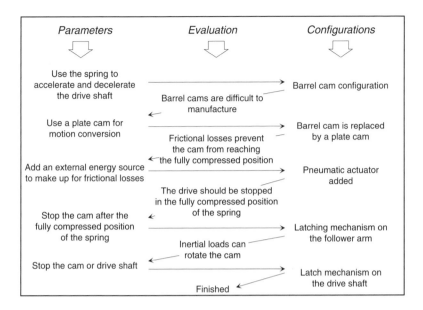

Figure 8.17 Summary of the parameter analysis process for conceptual design II.

8.7 Discussion and Summary

This case study demonstrates the strong link between need analysis and conceptual design. A good need analysis provides thorough understanding of the design task and valuable insights. The need for a flexible yet inexpensive drive system dominated the first conceptual design, which replaced the electric motor and speed reducer of conventional indexers with a pneumatic cylinder. Realizing that the net energy required was very small became the dominant parameter in the second conceptual design. Another critical issue identified in the need analysis was cost, and both designs have been evaluated and modified several times to address this requirement.

A prototype of the first design was built using two solenoids to simulate the forward and return strokes of a pneumatic actuator, as shown in Fig. 8.18. If the force from the actuator is assumed to be constant, then the torque on the drive shaft is also constant. The constant torque profile does not result in the modified sinusoidal

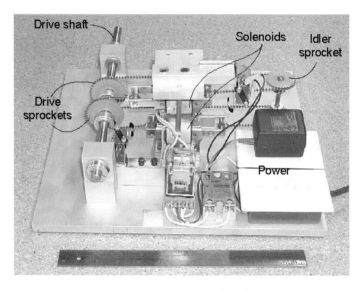

Figure 8.18 Prototype of the drive system with solenoids instead of a pneumatic actuator. (A 12-in. ruler is shown for scale.)

Figure 8.19 Prototype of the drive system in the fully extended position of the actuator. (The spring cannot be seen in the figure.)

acceleration profile. The constant torque causes a jerk during the initial motion of the load. However, the chain acts as a damper and reduces this jerk considerably. Even though the design does not fully emulate the modified sinusoidal acceleration profile, it satisfies the key requirements of cost and flexibility.

A prototype of the second design is shown in Fig. 8.19. It was tested to determine how closely it resembled the modified sinusoidal profile. Based on the results, the cam profile was slightly modified to accommodate the force profile of a typical pneumatic actuator.

8.8 Thought Questions

1. Critically evaluate the two conceptual designs presented in this chapter to identify their weaknesses. Then, apply parameter analysis to address these weaknesses and improve the designs.

2. Develop a conceptual design that incorporates a small electric motor instead of the pneumatic actuator in conceptual design II. Discuss the advantages and disadvantages of such a system.

3. Develop a design for powering a small boat. The user pushes two pedals alternately with his or her feet. This motion must be converted to continuous, unidirectional rotation of a paddle wheel. Perform detailed calculations to study the feasibility of the solution. Examine if a one-way clutch can be used to solve this problem.

8.9 Bibliography

An overview of cam design can be found in:

Norton, R. L. *Design of Machinery: An Introduction to the Synthesis and Analysis of Mechanisms of Machines.* 2nd ed. New York: McGraw-Hill, 1999.

The following book by the founder of Ferguson Machine Company describes various cam mechanisms, different acceleration profiles, and their advantages:

Neklutin, C. N. *Mechanisms and Cams for Automatic Machines: Influence of Dynamic Forces on High-Speed Machine Components.* New York: American Elsevier Publishing Company, 1969.

9

Conceptual Design Case Study: Equal-Channel-Angular-Extrusion Metalworking

A novel extrusion process with its unique machinery is the topic of this relatively complex case study. It begins with a general need for a method to improve material properties by deformation processing. The end result is quite a detailed design of a machine to carry out Equal-Channel-Angular-Extrusion metalworking. Metal bars are forced in this process through a bent channel and thus undergo strain hardening while preserving their size and shape. The benefits of this process are also explained.

9.1 The Initial Need

Plastic deformation in metal forming is used for two purposes: to change size and shape, and to improve material properties such as ductility, strength, and toughness. Forging, rolling, and extrusion are examples of forming and shaping processes in which adding strain to a workpiece results in refining the grain size, making the microstructure more uniform, and work hardening of the metal. Work hardening and finer grains are responsible for an increase in the shear strength, and hence the overall strength, of the material. This increase results from entanglements and impediments of dislocations by shear deformation, which usually takes place at 45° to the direction of the applied load. The greater the strain induced in the material during processing, the more the entanglements, and hence

157

a higher strength is obtained. Metalworking can also remove metal-
lurgical casting defects, such as cavities and voids, and produce
desired textures.

Adding strain is used extensively in metalworking processes—
for example, strengthening wire by reducing its cross-section by
drawing it through a die, forging the head of bolts, and producing
sheet metal for automobile bodies and aircraft fuselages by rolling.
Some of the common deficiencies of conventional processes can be
demonstrated by examining the process of extrusion. In extrusion,
material is forced through a die, which is similar to squeezing tooth-
paste from a tube. The three main problems are:

1. Material properties may not be homogeneous throughout
   the cross-section, due mostly to the nonuniform flow of the
   metal through the die.
2. Very large reductions of size are required to obtain significant work hardening or very fine and uniform grain size. If
   the dimensions of the final workpiece are to be relatively
   large, then the initial billet must be huge.
3. To produce the large changes in cross-section, high pressures
   need to be applied to the billet, which in turn result in high
   overall loads requiring large and expensive machines.

Figure 9.1 is a schematic of a conventional extrusion process. The
force F exerted on the punch translates into pressure p ($p = F/A_0$) on
the billet and forces it to change its cross-sectional area from A_0 to A_f.
The true strain, ε, experienced by the material is defined as:

$$\varepsilon = \ln(A_0/A_f) \tag{9.1}$$

The punch pressure required to produce ε while overcoming the
friction between the billet and the die is estimated in many textbooks from graphs and equations such as:

$$p = k\varepsilon \tag{9.2}$$

where k is an extrusion constant that depends on the work material,
the temperature, the die shape, and other factors. For example, to

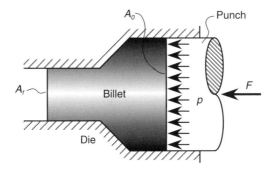

Figure 9.1 Schematic of conventional extrusion.

guarantee high-quality products from aluminum-based alloys, the ratio of initial to final cross-sectional areas (also called the extrusion ratio) should reach about 50 to 1 or, from Eq. (9.1), $\varepsilon \approx 4$. If a 100-mm diameter rod is desired, then the initial diameter should be about 700 mm. The required punch pressure (using a typical hot-deformation extrusion constant of $k \approx 75$ MPa at 350°C for aluminum alloys) will be $p = 300$ MPa from Eq. (9.2), and the corresponding punch load can reach about 12,000 tons. A press of such capacity will be large and expensive.

It follows that when improved properties through high strains are sought, the size of the finished product may be limited. The problem with the conventional processes is therefore their need to reduce the workpiece section considerably to produce a significant level of strain. What is sought in the present design is a new method to deform workpieces and subject them to high strain intensities without inducing dimensional changes.

9.2 Abbreviated Need Analysis

Performance

It is desired to process relatively small workpieces, such as circular- and rectangular-section rods with diameter or side of up to 100 mm, and length-to-diameter or side ratio of up to 10 to 1. The workpieces will typically be cast ingots, so high dimensional accuracy should not be expected. The machine should be capable of processing various metals, such as aluminum, copper, titanium, and their alloys, producing same-size products. The speed of processing is not critical because the process is intended for limited production volumes of selected alloys. From typical strain rates for the relevant materials, it seems that a production rate of 10,000 cm³/min. would be adequate. (This corresponds to one large $10 \times 10 \times 100$-cm billet every minute.)

The design should facilitate processing at various elevated temperatures, adjustable up to about 1000°C. To prevent the chilling effect (cooling of the workpiece upon contact with the tool), billet and die temperatures should be maintained to within ±50°C. Loading and unloading the machine can be done manually, due to both the limited size and weight of the workpieces and the low production rate.

Value

The value of the design is that of producing high-quality workpieces while eliminating the problem of size reduction associated with conventional processes. This means that the mechanical properties attainable with the new machine can be acquired by workpieces that are much larger than before. Eliminating the need for size reduction also has the potential of utilizing a much smaller and cheaper machine and saving energy in the processing of the materials. The value of the design will be even higher if it can utilize standard metalworking equipment.

Size

The machine may be quite large and heavy, comparable to high-capacity industrial presses used in processes such as forging and extrusion. The only size/weight limitations may be related to allowable loads on reinforced-concrete floors and fitting through large doors. Transportation constraints may also dictate a modular design.

Safety

The same precautions used with industrial presses apply to the current design. Hot parts should be shielded, and access to moving parts should be prevented during operation. Conventional safety techniques, such as dead-man controls, should be used.

9.3 Some Design Requirements

1. Deform metallic workpieces to induce high strains while maintaining the original shape. Typical materials to be processed are aluminum-based alloys such as 2090 aluminum–lithium. These have a flow stress σ_f of about 50 MPa at 350°C and require working to effective strains of about $\varepsilon \approx 4$.
2. Accept mainly rods with rectangular cross-sections. Circular and other sections may be desirable, too.
3. Accept rod diameter or sides of up to 100 mm, and length to diameter or side ratio of up to 10 to 1.
4. Adjust processing speed up to 100 mm/s.
5. Use hot deformation at variable temperatures up to 1000°C. Workpiece and tool temperature should be maintained within ± 50°C of the set temperature.
6. Loading/unloading may be manual.

9.4 Technology Identification

The processes mentioned before—namely, extrusion, rolling, and forging—represent different forming methods that use plastic deformation to produce shape changes and incorporate strain for work hardening or grain refinement. Any one of these processes can serve as a starting point for conceptual design, with the objective of keeping the strain-inducing aspect while preserving the workpiece size. For the present example, extrusion is chosen as the starting point.

9.5 Parameter Analysis

PI: The main deformation mechanism in plastic forming is shear. Try to accomplish shear deformation without reduction in the

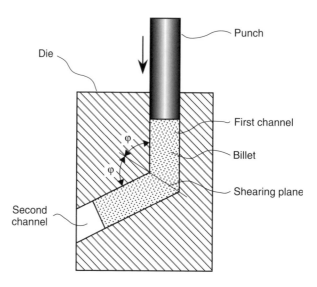

Figure 9.2 Extruding a billet through two intersecting channels of equal cross-section.

cross-sectional area by forcing a billet through a bent channel. The resulting simple shearing action will produce large strain, but the overall billet size and shape will not change.

CS: Figure 9.2 is a schematic of the process. The tool is a block with two intersecting channels of identical cross-section with an obtuse angle 2φ between them. The billet is a bar or rod, and it fits snugly in the channels. The process begins with the well-lubricated billet being placed in the first channel. A punch forces it around the corner and into the second channel. The deformation is realized by simple shearing in thin layers at the crossing plane of the channels. In this way, all sections of the billet (except small end parts) are deformed in the same manner. Finally, the punch moves back, and the deformed billet is withdrawn from the second channel.

Analysis of the velocity vectors of the material elements while they go through the intersecting channels shows that the effective strain experienced by the billet depends mainly on the bend angle and is given by:

$$\varepsilon \cong (2/\sqrt{3}) \cot \varphi \qquad (9.3)$$

which translates into the following sample values:

Bend angle, 2φ	φ	Effective strain, ε
180°	90°	0
150°	75°	0.31
120°	60°	0.67
90°	45°	1.15

The required punch pressure (neglecting friction between work-piece and die) can be found by energy-conservation analysis to be:

$$p = \sigma_f \varepsilon = (2/\sqrt{3})\sigma_f \cot \varphi \qquad (9.4)$$

and the punch force F is found from the punch pressure times the (constant) billet cross-sectional area, A:

$$F = p\,A \qquad (9.5)$$

For the case of extruding the aluminum-based alloy discussed in the need analysis, if the desired effective strain is about $\varepsilon = 4$ (corresponding to a 50-fold area reduction in conventional extrusion), we need to pass the billet four times through a 90° bend, or six times through a 120° bend, and so on. Because the shape and dimensions of the workpiece do not change, multiple passes can be done on the same machine. Moreover, if the extrusion takes place at an elevated temperature of 350°C, the billet flow stress, and thus the punch load will remain almost unchanged even after several passes.

The required punch pressure at each pass through a 90° bend (without accounting for friction) is found from Eq. (9.4) to be about 60 MPa, and the corresponding force on a 100-mm diameter billet is about 50 tons from Eq. (9.5), well within the capacity of small presses. Even though we neglected friction in the last calculation, it is interesting to note that our earlier calculation showed that to produce the same level of strain with conventional extrusion, 12,000 tons would be required!

E: The multiple-pass concept has another advantage: billet orientation can be controlled between passes. For some applications requiring very homogeneous properties, the billet should be rotated 180° after each pass, and the total number of passes should be even. Other applications may benefit from specific microstructures and textures. Laminar and fibrous structures would result from maintaining the same billet orientation during consecutive passes.

As implied by the table above, there is a trade-off between the number of passes required to obtain a desired effective strain and the angle 2φ between the channels. What is the optimal combination?

PI: The smallest possible angle between the channels is 90°. (Note that this is the largest possible angle through which the material can be deformed.) This angle would minimize the number of passes required to obtain the desired effective strain and simplify the design of the machine: the dies may even be installed on a conventional vertical press. On the other hand, it would require a larger punch force. Estimate the punch force including friction, and if it is not prohibitively high, choose 90°.

CS: Figure 9.3 shows a 90° setup. The length of the first channel is *L,* and the billet diameter is *d.* The punch pressure *p* in Eq. (9.4) is what is required for plastic deformation. But due to Poisson's effect, a billet under longitudinal compression would also push sideways, producing normal forces on the walls of the first channel, which result in a friction force. To overcome the effect of friction, the pressure exerted by the punch should be increased by Δ*p.* For static equilibrium, the added punch pressure times the cross-sectional area of the billet should equal the friction force on the walls of the first channel. The friction force is not uniform along the length of the first channel, but as a first approximation we shall assume that it is. The friction force is then found from the coefficient of plastic friction (μ) times the normal force. The normal force can be expressed as the hydrostatic flow stress in the billet times the area of contact between billet and channel:

$$\Delta p(\pi d^2/4) = \mu \sigma_f \pi dL \qquad (9.6)$$

or rearranging,

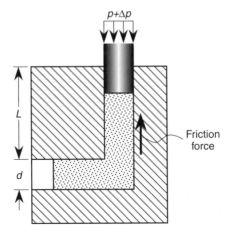

Figure 9.3 A right-angle extrusion die. The friction between billet and die requires additional punch pressure Δp.

$$\Delta p = 4\mu\sigma_f \, (L/d) \qquad\qquad (9.7)$$

A similar derivation can be done for other sections. For example, the added pressure to overcome friction with a square section of side b is

$$\Delta p = 4\mu\sigma_f \, (L/b) \qquad\qquad (9.8)$$

Using a good extrusion lubricant such as graphite, with $\mu \approx 0.1$, and assuming an L/d or L/b of 10, we get for aluminum-based alloys ($\sigma_f = 50$ MPa at 350°C) an added pressure of 200 MPa. Adding this to the 60 MPa calculated previously, we find that the total punch pressure comes up to 260 MPa. For a 100-mm diameter billet this translates to a punch force of about 200 tons, and for a 100-mm square billet the punch force is about 260 tons.

E: The previous calculation was pessimistic: if $d = 100$ mm and $L/d = 10$, it is implied that the length of the first channel is $L = 1$ m. Clearly, friction can be reduced considerably by making L shorter, for example, two to three times the diameter d. But even with the worst-case assumption, the 200- to 260-ton total punch force is still reasonable. Therefore, the 90° design is kept.

Shortening L to reduce friction seems like a good idea, but it presents two problems: possible buckling of the billet and the large compressive stress in it. The latter is very serious: as the preceding calculation shows, the punch pressure and the resulting compressive stress in the billet can be as much as five times the hot-deformation flow stress of the material. Even if friction is reduced significantly, the compressive stress will always exceed the flow stress and can cause the billet material to behave plastically outside the die.

PI: The billet must be fully contained in the die to prevent plastic deformation of the workpiece before entering the first channel. Obviously this would also prevent the billet from buckling.

CS: The structure of Fig. 9.3 is made to contain the full length of a billet whose length-to-diameter ratio is 10 to 1. The first channel needs to be about 1-m long to accommodate the longest billets. The second channel need not be that long; two to three times the billet diameter or side dimension seems sufficient to assure that the billet emerges straight from the die and does not "curl" upward. The punch, however, also needs to be about 1-m long so that it will reach the bottom of the first channel.

E: The punch is slender (with a length-to-diameter ratio as high as 10 to 1) and may buckle. But the punch could easily be supported along its length by some external, possibly retractable, braces, thus minimizing the free length and the danger of buckling. We shall ignore the issue of punch stability for now because the details of the die and punch design are still unclear.

The die in Fig. 9.3 seems difficult to manufacture because of the internally intersecting channels, which may have various cross-sections. Moreover, the design needs to accommodate different size and shape billets, so making new dies every time may be expensive.

PI: Reduce turnaround time and cost and simplify manufacturing by using interchangeable inserts for the channels, and one universal die housing.

CS: Figure 9.4 shows a design consisting of a split housing in which two inserts are fitted. The inserts are square outside and have the desired channel/billet cross-section inside (round section is

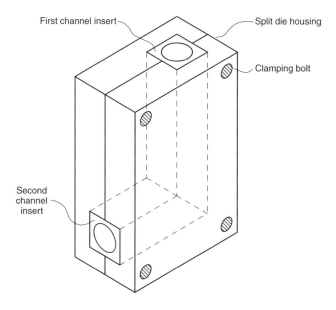

First channel insert

Split die housing

Clamping bolt

Second channel insert

Figure 9.4 Extrusion channel inserts clamped between two die halves.

shown). Several bolts can be used to clamp together the housing halves and secure the inserts in place.

E: Machining of the housing halves with their sharp internal corners may still be quite expensive. Moreover, if the design of Fig. 9.4 is sized to accommodate up to 1-m long billets but is used with much shorter workpieces, the process (placing the billet in the first channel, operating the press, and extracting the billet from the second channel) will be inconvenient.

PI: Extend the modular structuring principle so that simple building blocks can be combined to form a die for billets of different lengths.

CS: Figure 9.5 shows a modular structure made of five plates that can be bolted together. Channel inserts are no longer necessary because the sides of the plates form the extrusion channels. The design shown can accommodate billets with rectangular sections. The internal plates can be modified to fit billets with round or other sections.

E: The structure is now relatively inexpensive to make. It will also allow disassembly and extraction of a billet in case something

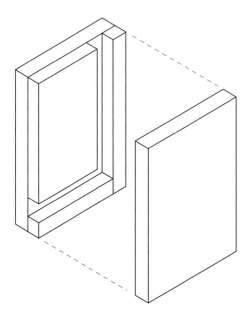

Figure 9.5 A modular die consisting of five plates and no inserts.

goes wrong during extrusion and a workpiece becomes stuck inside the die. Another advantage of the modular plates structure is the ease of adapting it to extrusion through angles other than 90° between the channels. Figure 9.6 shows how such a change takes place by modifying two of the three internal plates.

Because the billet is fully contained in the first channel, friction forces are very high (as calculated earlier). This will require a higher-capacity press but, even more important, may cause nonuniform deformation across the billet section. The friction may result in different strains being produced at the surface than at the center of the workpiece. This is a well-known problem with extrusion and is present even when the best available lubricants are used.

PI: As established earlier, the billet must be fully contained within the first channel. However, it may not have to *rub* against all the surfaces of that channel, if the channel is made to move with the billet. In this way, "bad" friction between the billet and channel may be replaced by friction between the moving chan-

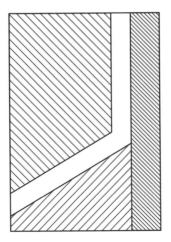

Figure 9.6 Two internal plates can be modified to produce angles other than 90° between the channels.

nel and the die housing which would not damage the work metal.

CS: Figure 9.7 shows a design in which the first channel consists of three "walls" that move with the square-section billet, and only one wall is stationary. (Having all four walls move with the billet would make the transition into the second channel difficult, perhaps impossible, to implement.) The sliding walls all belong to the same part—the slider—which can translate downward through an opening in the bottom of the die assembly. This slider rubs against stationary components of the die, but because the compressibly stressed billet is contained within it, the normal forces exerted on the two side plates cancel each other. The friction at the third rubbing surface (opposite the stationary wall) is between the slider and the die, so it can be controlled easily. (The large rubbing surface is better at preventing lubrication breakdown.) Only one surface experiences the moving billet–stationary wall friction.

An implementation for nonrectangular billet sections may be similar in nature, though different in some detail. As an important by-product, the punch can now be made much larger than before, having a section identical to that of the combined

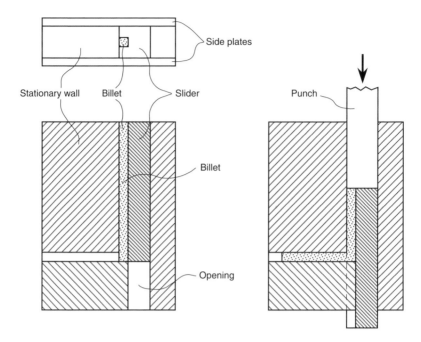

Figure 9.7 The slider constitutes three out of the four walls of the first channel.

billet and slider. This would eliminate the risk of punch buckling, which was identified earlier.

E: Although the total punch pressure required to overcome friction
     may not have changed much, the last configuration succeeds in
     significantly reducing the direct friction on the billet, thus greatly
     improving the uniformity of deformation. However, another
     known problem with extrusion is that brittle materials tend to
     suffer from defects and even fractures if no back pressure, or
     counterpressure, is applied during the process. The back pressure
     required to avoid these problems may reach about one-half the
     flow stress when extruding difficult-to-deform materials.

PI: Back pressure can be applied by introducing resistance in the
     outlet of the second channel. This would increase the required
     punch pressure: first, to overcome the back pressure, and sec-
     ond, because of the increased friction in the second channel
     due to the compressive loading on the billet. Experience with
     conventional extrusion shows that the back pressure needs to
     be proportional to the applied punch load but should not

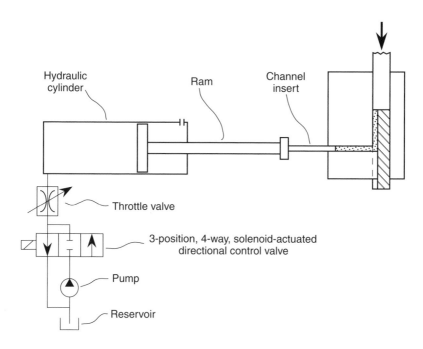

Figure 9.8 Back pressure in the second extrusion channel results from oil being forced out of the cylinder through an orifice.

exceed the flow stress of the work metal. It also needs to be easily adjustable for different materials.

CS: A single-acting vented hydraulic cylinder with a ram connected to an insert that fits in the second channel may be used for back pressure, as shown schematically in Fig. 9.8. The amount of back pressure can be regulated by using the proper valve and control system. The figure shows a variable-resistance orifice used to drain the oil from the high-pressure side of the cylinder.

To estimate the size of the required cylinder, let us assume a back pressure of one-half the flow stress, that is, 25 MPa for aluminum-based alloys at 350°C. When this pressure is applied by a 100 × 100-mm channel insert, it translates to a force of 250 kN. Producing this force at the other end of the ram using a standard 3000-psi hydraulic system will require a cylinder with an internal diameter of 125 mm.

E: A 125-mm diameter cylinder, its associated hydraulic components, and the control system will be bulky and expensive. A larger hydraulic cylinder may be required for materials that are more difficult to deform. An even more serious problem with the arrangement of Fig. 9.8 is its inflexibility: dimensional changes in the die (for example, to extrude shorter workpieces) may entail repositioning of the cylinder, such as mounting it higher or lower. Another, simpler method of providing back pressure may be beneficial.

PI: One of the simplest ways to produce proportional perpendicular forces is pressing down on an object that rests on an inclined plane. If the motion of the object were constrained to be vertical, the inclined base would move sideways. The downward force exerted by the punch may be converted to a horizontally acting back pressure by a wedge-like mechanical arrangement.

CS: Figure 9.9 shows an implementation wherein the bottom of the second channel is defined by the top surface of a plate—the horizontal slider—which moves with the billet. The slider rests on a stationary base tilted at an angle α to the horizon. During extrusion, the slider is pushed to the left by the action of the billet on the obstructing step, which is built into the end of the slider. The leftward motion of the slider is restricted by the inclined base, so back pressure is generated. The more punch pressure is applied and transmitted to the obstructing step by the billet, the greater the normal force on the inclined base. A greater normal force causes a larger friction force, resulting in the back pressure increasing in proportion to the punch pressure. The alignment of the slider is maintained by two sets of horizontal tongue-and-groove connections to the side plates. During operation, the whole die assembly also moves slightly upward—a distance marked by x in Fig. 9.9—on four guide pins provided at the bottom of the assembly.

Because the horizontal slider needs to accommodate the full length of the billet, a change is made to the structure of the first channel. The first channel now consists of two (instead of three) sliding walls and two stationary walls. Accordingly, the single vertical slider of Fig. 9.7 is now split into two rectangular-section sliders, each constituting one channel wall. The punch now assumes an H-shaped cross-section, as shown in Fig. 9.10. The reduced friction in the second channel due to the horizontal slider partly makes up for the added friction in the first channel.

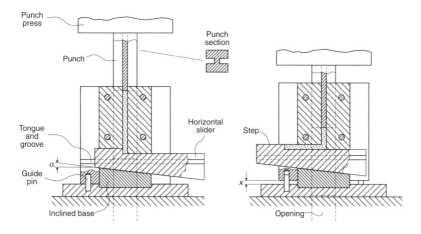

Figure 9.9 Back pressure is induced in the second channel by the wedge effect of the horizontal slider on the inclined base.

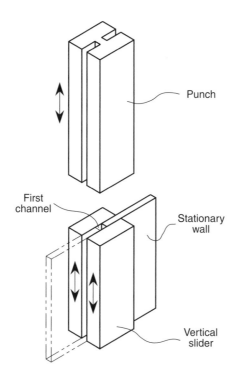

Figure 9.10 The walls of the first channel are created by the two vertical sliders and two stationary walls. The H-shaped punch is also shown.

The base of the die assembly still contains an opening to allow the vertical sliders to descend during extrusion.

E: This seems like a good arrangement, but how will the horizontal slider return to its place at the end of each extrusion process? Gravity alone may not do the job if the current die assembly is to operate with existing horizontal presses; that is, the extrusion is done in a horizontal plane.

PI: Use springs to return components to their initial position.

CS: The die assembly, which rises during extrusion (through distance x in Fig. 9.9), could be returned by springs. Figure 9.11 shows how each of the four vertical guide pins of Fig. 9.9 is replaced by a special spring-loaded screw, which in turn requires a cavity in the punch-press bed. During the upward motion of the die assembly, the springs will compress. When the punch force is removed, the springs will extend and pull the die assembly back down. This arrangement will work even if the die is placed on its side and gravity does not assist the return motion.

E: Friction present between the moving parts may interfere with the spring action and prevent the die parts from being repositioned accurately. Moreover, the vertical sliders also need to be repositioned at the end of their stroke.

PI: Use hydraulic actuators to reposition parts. The horizontal slider will be returned by new cylinders. The vertical sliders can use the punch press itself for the return motion if a joint is designed to allow pulling the slider up by the punch.

CS: To return the horizontal slider sideways and downward along the inclined plane of the base, a pair of hydraulic cylinders is added with a cross-bar connected to the slider, as shown in Fig. 9.12. These cylinders can be quite small and can be incorporated into the main hydraulic system of the press. In contrast to the hydraulic cylinder of Fig. 9.8, which was mounted independently of the die, the current pair of cylinders is attached to the die, so they all move up and down in unison during extrusion. The springs proposed for incorporation into the base of the die assembly (Fig. 9.11) are no longer required, so the original guide pins (Fig. 9.9) can be used instead.

To return the vertical sliders upward, a hook-type connection is implemented between the punch and the sliders. As shown in

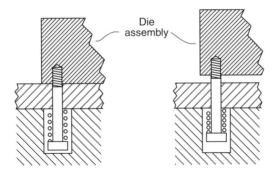

Figure 9.11 Spring-loaded screws replace the guide pins to reposition the rising die assembly.

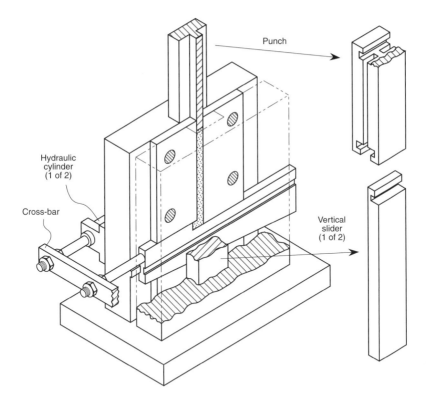

Figure 9.12 Hydraulic cylinders and tongue-and-groove connections at both ends of the punch are added to facilitate repositioning.

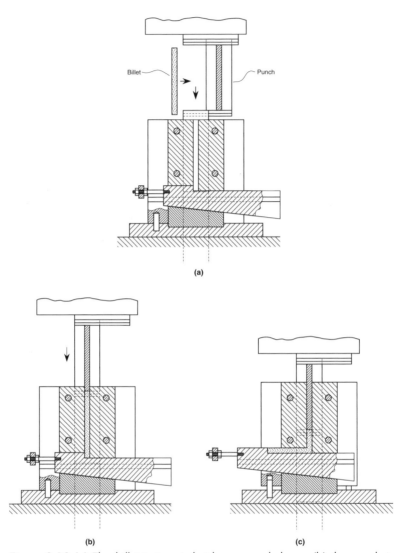

Billet

Punch

(a)

(b)

(c)

Figure 9.13 (a) The billet is inserted sideways and down; (b) the punch is moved left and extrusion begins; (c) about two-thirds of the stroke is completed.

Fig. 9.13, the operator moves the punch sideways on its tongue-and-groove connection to the punch press to allow placement of the billet in the first channel. The punch is then returned while engaging it with the vertical sliders. During extrusion, the punch presses down on the sliders and billet, and pulls the sliders up at the end of the stroke.

E: The horizontal cylinders can serve two additional purposes: they can be used for supplemental back pressure, and they can help retract the billet at the end of the extrusion process. They only need to be of the double-acting variety to perform all three tasks. Billet extraction, however, may still be a problem. Even if the horizontal slider is made quite long, there is no guarantee that the billet would emerge from the second channel when the cylinders are made to pull on the slider. In fact, the billet may remain "stuck" between the three stationary walls of the second channel.

PI: Create an attachment between the billet and the horizontal slider so that when the slider is pulled by the hydraulic cylinders, the billet will follow it.

CS: A small cavity created in the horizontal slider will be filled by the work metal when it is in a plastic state, thus creating a "hook" for the slider to pull the billet. This is shown in Fig. 9.14.

E: The billet will have a small protrusion at its end, created by the cavity in the horizontal slider. This protrusion will have to be ground off before inserting the billet for another extrusion pass. This solution is acceptable because the very ends of the billet anyway do not undergo the shearing action of angular extrusion and do not possess the improved material properties of the rest of the billet. Still missing from the design is a provision for maintaining the billet at constant temperature during extrusion.

PI: The extrusion process is quite fast: extrusion speeds of up to 100 mm/s can be applied by the current design when coupled with conventional presses. The billet needs to be preheated to the desired temperature before entering the die. The die should also be heated to prevent the well-known chilling effect of the billet material when contacting the die.

CS: An induction furnace to heat billets will be included in the system. This is a standard, off-the-shelf item and need not be designed. A heater will be added around the die, and the die assembly may also be separated from the punch-press bed by an insulating layer. Figure 9.15 offers an overview of the design with the added heater. A control system will also be included to monitor the die temperature and regulate the heater operation accordingly.

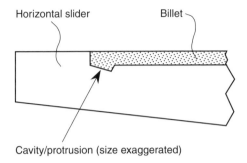

Figure 9.14 A cavity at the front end of the horizontal slider creates a "hook" in the billet to help with its extraction.

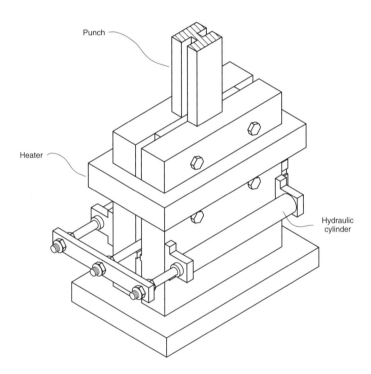

Figure 9.15 A heater is mounted around the die assembly to maintain constant temperature.

E: All major issues seem to have been addressed. The device has been designed to operate with vertical presses; however, it can be adapted to horizontal operation. The final conceptual design applies to processing of rectangular-section billets. Some adap-

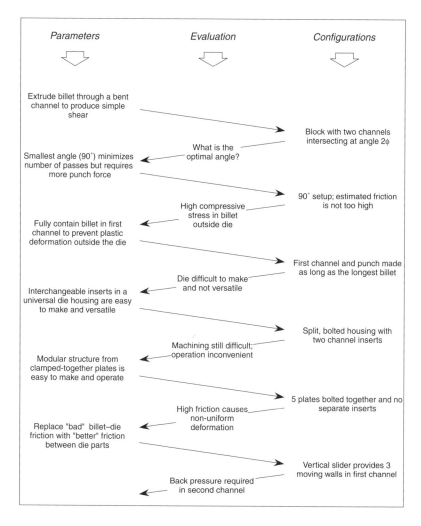

Figure 9.16 Summary of the ECAE parameter analysis process. *(Continues on page 180.)*

tation may be required to accommodate circular and other sections, although the overall concept is still valid.

9.6 Discussion and Summary

This chapter presented a relatively complex design example whose parameter analysis process is summarized in Fig. 9.16. It begins with

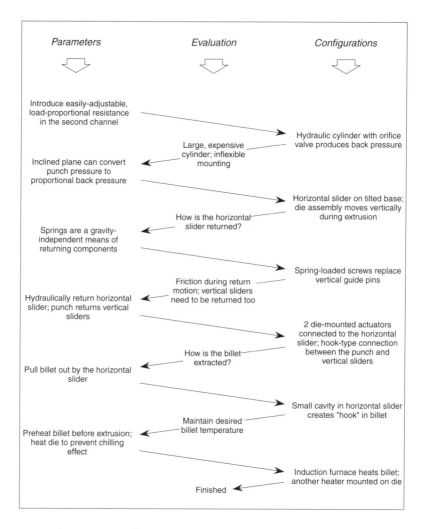

Figure 9.16 Continued

an abstract need to devise a method for improving material properties by deformation processing, and it ends with a relatively detailed design of an Equal-Channel-Angular-Extrusion (ECAE) metalworking machine. Accordingly, some of the identified parameters represent innovative breakthroughs, while others deal mainly with practical aspects of structuring a viable machine.

The next design phase will add a control system for the extrusion machine, including means of activating the horizontal cylinders and sensors (e.g., micro-switches) to signal the end of the vertical and horizontal strokes. Some heat-transfer calculations are required to size the induction furnace and the die heater. Shields may also be added around the die assembly to prevent the operator from accidentally touching the die heater or any of the moving parts.

9.7 Thought Questions

1. The provision made in the design of this chapter to extract the billet from the second channel (i.e., the cavity in the horizontal slider) is not a very elegant solution. Design a better method for billet extraction to be incorporated in the design.

2. The advantages of using sliders, or channel walls that move with the billet, became clear during the design process. Design a die assembly that has at least three "moving walls" in the first channel, similar to the concept of Fig. 9.7. Can a die be designed to have four moving walls in the first channel?

3. How would you turn angular extrusion into a continuous process, suitable for handling very long workpieces (e.g., metal wire)? It would be impractical to enclose very long billets in the first channel, so the conceptual solution should be very different from the design presented above. One direction worth exploring is an adaptation of the Conform extrusion process, where the billet is pulled, rather then pushed, into the die by a rotating wheel.

4. One common practice with conventional extrusion is using a "dummy" billet to help extract the workpiece from the die. Use this concept for billet extraction and modify the design of this chapter accordingly.

5. Instead of using a dummy billet, workpieces could be fed continuously so that each billet pushes its predecessor out of the second channel. However, the design needs to be modified; for example, another horizontal slider should take the place of the first hori-

zontal slider when the latter has been displaced by a preceding billet. Make the necessary provisions to the design to accommodate this semicontinuous process.

9.8 Bibliography

Some reference material may be useful in better understanding this chapter. General books on manufacturing processes usually have chapters on material behavior and fundamentals of extrusion. Two such books are:

> Groover, M. P. *Fundamentals of Modern Manufacturing: Materials, Processes, and Systems.* Upper Saddle River, NJ: Prentice-Hall, 1996.
>
> Kalpakjian, S. *Manufacturing Engineering and Technology.* 2nd ed. Reading, MA: Addison-Wesley, 1992.

Numerous books are available on more specific topics related to this chapter. Chapter 4 in the following book contains information on the Conform extrusion process, which may be useful for thought question number 3:

> Blazynski, T. Z. (ed.) *Design of Tools for Deformation Processes.* New York: Elsevier Applied Science Publishers, 1986.

Harris presents a more theoretical treatment of the behavior of metals in several manufacturing processes:

> Harris, J. N. *Mechanical Working of Metals: Theory and Practice.* Elmsford, NY: Pergamon Press, 1983.

The design presented in this chapter is based to a large extent on the following patent:

> Segal, V., Goforth, R. E., and Hartwig, K. T. *Apparatus and Method for Deformation Processing of Metals, Ceramics, Plastics and Other Materials.* U.S. Patent 5,400,633, 1995.

Several papers describing Equal-Channel-Angular-Extrusion follow. The first is an overview of the process, and the others mostly demonstrate its benefits when applied to specific materials:

Segal, V. M. "Materials Processing by Simple Shear." *Materials Science and Engineering* A197, (1995): 157–164.

Segal, V. M., Goforth, R. E., and Hartwig, K. T. "The Application of Equal Channel Angular Extrusion to Produce Extraordinary Properties in Advanced Metallic Materials." *Proceedings of First International Conference on Processing Materials for Properties,* Hawaii, November 7–10, 1993, pp. 971–974.

Goforth, R. E., Segal, V. M., Hartwig, K. T., and Ferrasse, S. "Production of Submicron-Grained Structure in Aluminum 3003 by ECAE." *Proceedings of Superplasticity and Superplastic Forming Conference,* Las Vegas, February 13–15, 1995, pp. 25–32.

Semiatin, S. L., Segal, V. M., Goetz, R. L., Goforth, R. E., and Hartwig, K. T. "Workability of a Gamma Titanium Aluminide Alloy During Equal Channel Angular Extrusion." *Scripta Metallurgica et Materialia* 33, No. 4 (1995): 535–540.

10

Need Analysis and Conceptual Design Case Study: "Ball Mover"

A group of second-year engineering students applied the need analysis and parameter analysis methodologies to design an entry in the 1997 American Society of Mechanical Engineers (ASME) Student Design Contest. The objective of the competition was to automatically transport two ping-pong balls and one golf ball between two locations. This case study describes the valuable insights gained by these students during the need analysis stage and the implication of these insights for the conceptual design process. It also highlights the natural tendency of designers to subconsciously make implicit assumptions and not recognize them throughout the design process. These assumptions often lead to conceptual oversights and may cost the design its competitive edge.

10.1 The Initial Need

Most mechanical designers handle the challenge of transporting objects from one location to another quite easily. The 1997 ASME Student Design Contest provided an opportunity for student members to learn some of the intricacies of designing material handling systems. The objective of the competition was to design and build a device to safely transport two ping-pong balls and one golf ball from the top of a platform to a receiving box separated by a given distance, as shown in Fig. 10.1. Contest regulations allowed the com-

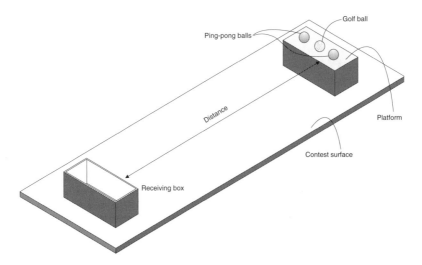

Figure 10.1 Setup for the 1997 ASME Student Design Contest.

petitors to choose their own distance anywhere between 50 and 211.8 cm. The score would be computed as

Score = Average velocity = d/T

where *d* is the distance in cm between the platform and the receiving box and *T* is the total time in seconds from the start signal to when the balls come to complete rest in the receiving box. Note that the goal was not to transport the balls in the shortest time over a set distance, but rather to transport balls at the highest *average velocity*. This means that the competitors could choose the desired distance (within the specified limits) for which the ratio of distance to time should be maximized. Key competition guidelines are summarized as follows:

1. The system should be powered by a single 1.5 V size AA alkaline battery and a single specified dc motor. The motor cannot be modified, and additional energy sources or motors are not permitted.
2. In all energy storage devices except for the battery, the energy stored at the end of the run should be greater than or equal to the energy at the start.

185

3. After the start of the run, any external communication, interaction, or influence is not allowed; that is, the system must be autonomous.
4. The complete system must fit in a 16 × 16 × 32-cm box before assembly or placement. If assembly is required, the system must be put together and ready for the run in five minutes. The assembled system can only contact the contest surface, the receiving box, and the platform.
5. The outside dimensions of both the platform and the box are 16 × 16 × 32 cm. They are constructed of 1/4″ plywood. The balls initially rest in 1/2″-diameter holes on the top of the platform, with 8 centimeters from center to center.
6. The system must deliver the three balls from the platform into the receiving box within one minute of the start signal.

Designing a device to transport the balls was relatively easy. However, the following need identification and analysis indicate that the real challenge lay in doing so at the highest average velocity.

10.2 Need Identification

The students used the black-box technique to precisely identify the need, as shown in Fig. 10.2. The balls on the platform, the electrical energy from the battery, and the start signal from the user were the three inputs to the black box. The desired output was the balls resting in the receiving box. The designed device had to transform the three inputs to the required output. The need was restated as "transport the balls from the platform to the box using the energy from the battery at the highest average velocity." Note that the students consciously avoided the statement "move the balls at the maximum velocity" because it related to only one part of the total time (total time = initial time before the motion starts + acceleration time + time of motion at constant velocity + deceleration time to full stop).

Some of the questions that came to the students' minds while defining the need were:

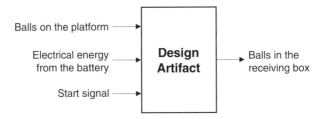

Figure 10.2 Black-box description of the design artifact.

- What is the ideal distance between the platform and the receiving box?
- Is it advantageous to go for the full distance?
- What is the ideal or theoretically highest possible score?
- What are the characteristics of an ideal solution?
- What are the important hurdles to achieving the highest score? What is preventing us from creating an ideal solution?

The students performed the following need analysis to answer these and other questions, gain deeper understanding of the design task, and generate the design requirements.

10.3 Need Analysis

Performance

The primary objective is to create a design that gives the highest score or performance. The score is a function of the distance (a variable that can be chosen to optimize the score) and the total time to transport the balls. As mentioned earlier, the total time can be divided into initial time before the motion starts, acceleration time, time of motion at constant velocity, and deceleration time to full stop.

For establishing targets, it is important to get a rough idea of the ideal score. To simplify the calculations at this stage, the deceleration time to full stop, and the distance traveled by the balls during deceleration are assumed to be zero. The transportation of the balls involves energy, so the initial calculation can use energy balance.

Energy from the battery can be converted to potential and kinetic energies of various components:

$$\eta E_{Battery} = PE + KE$$

where η is the combined efficiency of energy conversion (from electrical to mechanical) and transfer (from the battery to the motor shaft and from there to the balls and other moving parts), $E_{Battery}$ is the energy output from the battery, PE is the potential energy increase in various components, and KE is the kinetic energy given to any moving object. The energy output from the battery is given by

$$E_{Battery} = P \times t$$

where P is the power output from the battery and t is the duration of energy output. The kinetic energy gained by all the moving components is:

$$KE = (1/2) m_{MovingMass} v^2$$

where $m_{MovingMass}$ is the total mass of all moving objects and v is their instantaneous velocity. (At this point, it is assumed that all moving objects have the same velocity.) Combining the last three equations gives

$$\eta \times P \times t = PE + (1/2) m_{MovingMass} v^2$$

and rearranging the terms shows the instantaneous velocity to be

$$v = \sqrt{2 \times (\eta \times P \times t - PE)/m_{MovingMass}}$$

The last expression means that the instantaneous velocity can be increased by:

- Increasing the duration, t, of power output from the battery. Note that the maximum duration is equal to the total time for the run, T.

- Eliminating any potential energy increase. This can be done by not lifting any component from its original position.
- Decreasing the moving mass, $m_{MovingMass}$, by eliminating any motion except for that of the balls. In this ideal situation, the moving mass will be equal to the mass of the balls only.
- Increasing the efficiency of energy conversion and transfer, mostly by reducing frictional losses.

The instantaneous velocity is the velocity achieved by the balls after the energy is supplied for a certain duration of time. However, our interest is in determining the average velocity (d/T) and the optimal value for the distance (d). The average velocity will depend on whether the energy output from the battery is for a brief period at the start of motion or continuously throughout the run. Solutions that employ the former strategy will convert the electrical energy to mechanical energy and store it for a short period of time during which the balls do not move. The stored energy will then be released to increase the velocity of the balls.

The velocity profile in Fig. 10.3a shows the short idle period at the beginning of the run, the sharp increase in velocity as energy is supplied to the balls, and the gradual drop in velocity due to frictional losses. The area under this curve represents the distance traveled. As can be seen, increasing the time beyond the start of motion increases the distance covered. The average velocity profile (Fig. 10.3b) is calculated as the ratio of this distance to the total time. It shows that when the distance is short, the initial idle time contributes significantly to the total time, thus reducing the average velocity. Increasing the distance initially increases the average velocity, but beyond a certain point, frictional losses reduce it.

Increasing the initial idle time before the motion starts increases not only the peak velocity, but also the total time. Therefore, it adversely affects the final score. On the other hand, decreasing the initial time reduces the velocity (because there is less time for the electrical-mechanical energy conversion), so it also has a negative effect on the final score. Any solution that employs this strategy entails a trade-off between the initial idle time and the peak velocity, so this solution strategy is discarded.

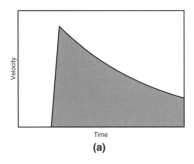
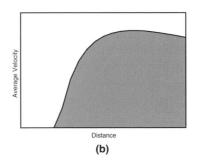

(a) **(b)**

Figure 10.3 (a) Instantaneous velocity profile and (b) average velocity profile for the initial energy output strategy.

The velocity profile for designs that employ the second solution strategy—that of continuously accelerating the balls—can be obtained by plotting the instantaneous velocity at the end of different time periods of energy transfer, as shown in Fig. 10.4a. The area under the curve is equal to the distance traveled by the balls, and this is shown in Fig. 10.4b. Because the balls cover greater distance in a shorter time interval during the later part of the run, increasing the distance between the platform and the receiving box will increase the average velocity. This insight is depicted graphically in Fig. 10.4c. Additional calculations also show that using this strategy, the balls can be moved from the platform to the receiving box within one second. A target of one second is therefore set for transporting the balls.

The understanding gained so far from the above analysis can be summarized as the following conclusions:

- Continuously accelerate the balls for an optimal score. In other words, keep drawing energy from the battery for the entire duration of the run.
- Use the maximum allowable distance.
- Reduce the moving mass.
- Avoid increasing potential energy of components.
- Increase the efficiency of energy conversion and transfer.

Value

The value of a successful design cannot be estimated, in this particular case, in monetary terms. The competition is a valuable experi-

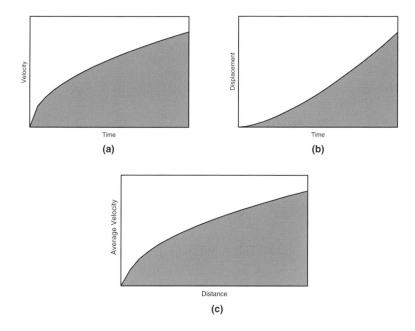

Figure 10.4 (a) Velocity, (b) displacement, and (c) average velocity profiles for the continuous acceleration strategy.

ence for students because it exposes them to various facets of the design process. Students can also evaluate their design skills when competing at the regional and national levels, and they take great pride in their achievements. From the cost viewpoint, a budget of $100 was allocated to purchase off-the-shelf components and materials.

Size

The competition rules specify that the complete system must fit in a $16 \times 16 \times 32$-cm box. The dimensions of the platform and the receiving box, the initial ball locations, and other geometrical data are as described in Section 10.1.

Safety

Although the design uses electrical energy, there is no electrical hazard. While the balls are moving, they might bounce uncontrollably onto the contest surface, resulting in an ineligible run. This rule simulates a drop from the material-handling system that would damage the product in real-world situations and might also be unsafe. Competition guidelines explicitly state that the balls cannot be damaged during the run.

Special

The device must meet the ASME Student Design Competition rules and regulations. Some requirements in this category include:

- Setup (assembly) time should be under 5 minutes, and all the team members may participate in it.
- The device can contact only the platform, the receiving box, and the contest surface—but not the balls—before the start of the run.
- There should be no residue left on the balls at the end of the run.

10.4 Design Requirements

Table 10.1 summarizes the requirements generated in each of the need analysis categories. The students not only identified the real need but also made important decisions. For instance, based on the calculations in the performance category, the target time for the complete run is set to 1 second, and the distance between the platform and the receiving box is set at 211.8 cm.

10.5 Technology Identification

In the need analysis stage, the designers identified the need for continuously moving the balls to achieve a high score. Three preliminary ideas for designs that continuously accelerate the balls are as follows:

1. *Drawbridge.* A drawbridge that connects the platform and receiving box can be raised at the platform end. The balls will roll down the drawbridge and into the receiving box due to gravity.

Table 10.1 The "ball-mover" requirements list.

Category	Design Requirements
Performance	1. The device must move two ping-pong balls and one golf ball in less than 1 minute. Target time is 1 second.
	2. The target distance between the platform and the box is 211.8 cm.
	3. Power source: One size AA alkaline battery (rated voltage: 1.5 V, rated current: 1 A).
	4. Motor: Radio Shack dc motor model 273–223 (no load speed: 5,700 rpm, max. efficiency: 45% at 4100 rpm, power used at max. efficiency: 0.39 W).
	5. Ping-pong ball: Mass: 2.5 g. Diameter: 38 mm (1.5 in.).
	6. Golf ball: Mass: 45 g. Diameter: 41 mm (1.6 in.).
	7. The energy stored in all energy storage devices (except for the battery) at the end of the run should be greater than or equal to the energy at the start.
	8. The system must be autonomous after the start; that is, external communication, interaction, or influence of any kind is not permitted.
Value	9. Cost of materials and purchased items should be under $100.
Size	10. The complete system must fit in a 16 × 16 × 32 cm box.
	11. Platform dimensions: 16 × 16 × 32 cm.
	12. Receiving box dimensions: 16 × 16 × 32 cm.
	13. The balls initially rest in 1/2" holes centered on the top of the platform 8 cm apart from center to center.
Safety	14. Balls or any other part of the mechanism should not bounce in an uncontrollable manner.
	15. Balls should not be damaged in any way.
Special	16. Setup (assembly) time should be under 5 minutes.
	17. Number of people available for setup: 4.
	18. The device can only contact the platform, the box, and the contest surface at the start of the run.
	19. The device should meet the 1997 ASME Student Design Contest rules and regulations.
	20. There should be no residue on the balls at the end of the run.

2. *Puller rod.* A fixed bridge can connect the platform with the receiving box. A cross-bar, or puller rod, moves the balls along the bridge and into the receiving box.

3. *Cable car.* A transport device will pick up the balls, carry them on a cable running overhead from the platform, and release the balls into the receiving box.

The first two concepts served as initial conditions for the two parameter analysis processes described below. Note how the drawbridge concept seems to contradict the "eliminate any potential energy increase" conclusion from the need analysis.

The final score in the competition also depends on the stopping time. Two concepts for bringing the balls to rest are:

1. *Damping.* The kinetic energy of the balls is gradually dissipated until they stop completely.
2. *Capturing.* The balls are arrested in a contraption that prevents their motion.

10.6 Parameter Analysis of Conceptual Design I

PI: Continuously move the balls using the drawbridge idea.

CS: An initial sketch for this concept is shown in Fig. 10.5. The motor mounted on a support tower lifts the drawbridge using a cable (or a fishline). The balls roll down the drawbridge into the receiving box. Holes or slots (not shown in the figure) can be machined in the drawbridge so that the balls do not contact it at the start of the run, as required by the contest regulations. The raised longitudinal edges of the drawbridge prevent the balls from falling off.

E: Several key problems with this design are:

- A significant portion of the energy goes into lifting the weight of the bridge.
- Because the support tower and the drawbridge must fit in the box, they should be assembled from smaller components. This problem seems to be easy to solve by modifying the design.
- There is no stopping mechanism for the balls.

The first problem is the most difficult because it is inherent to the drawbridge concept and can potentially render the whole idea nonviable.

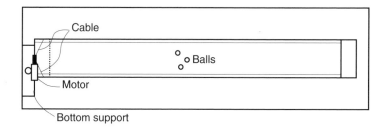

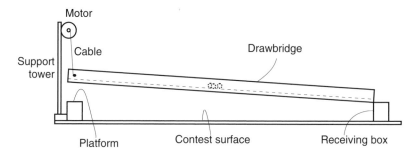

Figure 10.5 The initial configuration for the drawbridge idea to move the balls.

PI: The concept of counterbalancing is used in some designs, such as elevators, to reduce the required lifting force. Apply this idea to the drawbridge.

CS: The cable used to raise the drawbridge (see Fig. 10.5) is made to go over a pulley mounted on the motor shaft and attach to a counterweight at the other end. The weight of the drawbridge can thus be balanced, and the motor would not need to lift the weight of the drawbridge.

E: The motor must now overcome the inertia of not only the drawbridge, but also the counterweight. This will result in slower acceleration at the beginning of the run and adversely affect the total time and score. Based on this insight, the concept of counterbalancing the drawbridge is discarded.

PI: Decrease the energy required for lifting the drawbridge by reducing its weight.

CS: The drawbridge can be constructed of fiberglass fabric stiffened with epoxy. The composite sheet can be selectively stiffened so that it can be folded to meet the size requirement. It can also be shaped like a converging channel, or trough, for the balls to roll in.

195

E: The students built several models and realized that the design modifications were difficult and expensive to incorporate due to the required curing time of the composites and the cost of the materials. Even though the concept seemed very attractive, it was abandoned due to the implementation difficulties.

PI: Reduce the weight of the bridge by using a plain fabric.

CS: A lightweight fabric can be used to connect the platform and the receiving box. When the cloth is lifted by a fishline connected to its two corners at the platform end, the balls will roll down the slope. Because the fabric is flexible, it automatically forms a channel to guide the balls. A sketch of this idea is shown in Fig. 10.6.

E: This design effectively reduces the weight. The students performed a simple experiment wherein they lifted the fabric at the platform end using the motor. The balls rolled down the fabric and moved the full distance in about 8 seconds. The experiment revealed an important problem: The fabric sagged due to the weight of the golf ball, as shown in Fig. 10.7. As a result, the slope ahead of the balls sometimes became negative, and the balls decelerated in the final stages of the run.

PI: Prevent sagging by reducing the stretch of the fabric.

CS: The following modifications are incorporated into the design:

- The fabric is changed to a thin perforated plastic sheet. The plastic sheet stretches to a lesser extent under the golfball weight. An additional advantage of the perforated plastic sheet is that it reduces the rolling resistance and the air drag associated with the lifting.
- A thick plastic backup sheet is added. It connects the platform with the receiving box and is not raised during the run. It is firmly attached to prevent sagging around the receiving box end of the run, as shown in Fig. 10.8.

E: There is no mechanism to stop the balls when they arrive in the receiving box.

PI: Stop the balls by using the damping idea listed in the technology identification section.

CS: Several damping techniques to remove energy from the balls were tested. Students found that a tray containing very fine

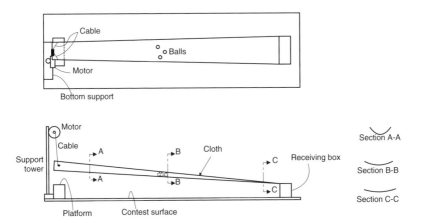

Figure 10.6 A cloth connects the platform with the receiving box, naturally forming a channel to guide the balls.

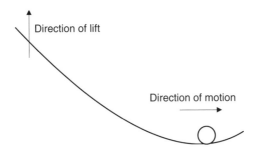

Figure 10.7 The sagging due to the golf ball creates a negative slope in front of the balls.

powder packed in thin plastic wrap was the most effective method for damping the motion. The plastic wrap prevented the balls from picking up any residue, which was one of the design requirements.

E: Several problems persisted throughout the development of this concept:

- The balls moved unpredictably in the initial stages of the run, when they first touched each other in the converging channel. Occasionally, they collided with each other and jumped out of the channel onto the contest surface.

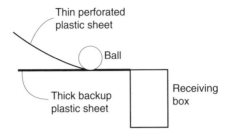

Figure 10.8 A plastic backup sheet reduces the sagging problem.

- The sagging problem still persisted around the middle of the span.
- To continuously accelerate the balls, the required height was more than 8 ft. The assembly time requirement and stability considerations limited the maximum height to about 6 ft.
- The balls took approximately 2 seconds to stop in the damping tray due primarily to their rolling motion. Depending on the material in the wrap, either the ping-pong balls or the golf ball took more time to come to complete stop.

Looking at these problems, it was decided to abandon the draw-bridge concept entirely. Important insights gained during the concept development are:

- Moving the balls individually provides better control of the balls. For instance, different damping methods can be applied to different balls.
- The rolling motion of the balls contributes significantly to the stopping time and thus to the total time.

Figure 10.9 shows a summary of this parameter analysis process. The decision to relinquish this concept, shown as a dead-end in the last evaluation step, led to backtracking all the way to the technology identification stage. Starting the second parameter analysis process with a different initial concept (but with some new insights) is described in the next section.

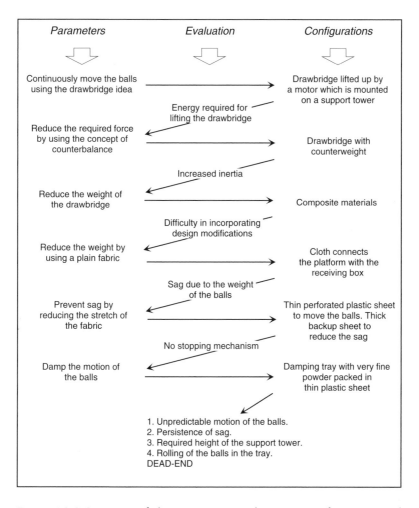

Figure 10.9 Summary of the parameter analysis process for conceptual design I.

10.7   Parameter Analysis of Conceptual Design II

PI: Move the balls individually from the platform to the receiving box using a puller rod.

CS: Three sets of tracks, one for each ball, are shown in Fig. 10.10. The tracks are firmly supported. The motor is mounted on the receiving box, and a fishline runs from the motor pulley along the middle tracks to the puller rod. The guides on the puller rod

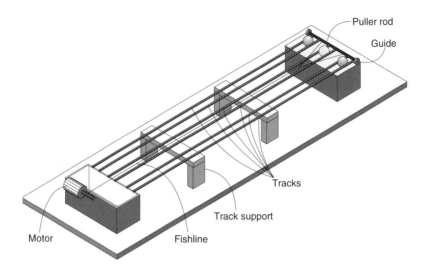

Figure 10.10 The initial configuration for the puller-rod concept.

prevent it from inadvertently leaving the tracks. When the motor turns, it rotates the pulley and winds the fishline. The fishline pulls the puller rod toward the receiving box and, in the process, rolls the balls along the track. To minimize the moving mass, the puller rod is made of thin aluminum tubing.

E: No stopping mechanism exists.

PI: One of the insights gained in the previous conceptual design process is that the rolling motion during the stopping phase contributes significantly to the total time. Based on this insight, rolling motion during the stopping of the balls should be prevented.

CS: A rather simple solution to address this parameter is shown in Fig. 10.11. At the end of the run, each ball falls into its own cavity. The shape of the cavities is determined by using the principle of kinematic constraints. It requires three noncollinear point contacts to fully constrain the ball from rolling. A trihedral cavity formed by three planar surfaces provides such contacts and constrains the ball. The balls come to a complete rest in less than 0.1 second.

E: The layout of the overall system indicates that it would function well. However, the system does not meet the size requirement.

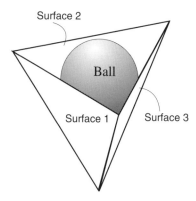

Figure 10.11 A ball rests in the trihedral cavity formed by three planar surfaces.

PI: Create modular subassemblies that fit in the $16 \times 16 \times 32$-cm box.

CS: Each track is assembled from 31-cm long, 1/4″-diameter aluminum tubing. Dowel pins are used to connect the sections together. The track supports are assembled from three wooden blocks. The trihedral cavity is created on-site by folding a cardboard sheet and fixing it in the wooden holder. All these modifications are shown in Fig. 10.12.

E: A small section of the track was built to identify potential problems. During the testing, the students noticed that the track rods moved in the transverse direction due to the weight of the golf ball.

PI: Prevent the transverse motion of the track rods.

CS: Slots are cut into the track rods, and dowel pins are inserted on top of the horizontal support. When assembled, the dowel pins fit into the slots in the tracks and prevent the transverse motion of the tracks, as shown in Fig. 10.13.

E: The puller rod does not stop at the end of the run.

PI: Stop the puller rod at the end of the run using mechanical stops.

CS: Two rods can be incorporated into the holder of the trihedral cavities. The rods protrude from the top of the receiving box and stop the puller rod.

E: When the motion was taped with a high-speed video camera, the puller rod was observed to lag behind the balls in the final

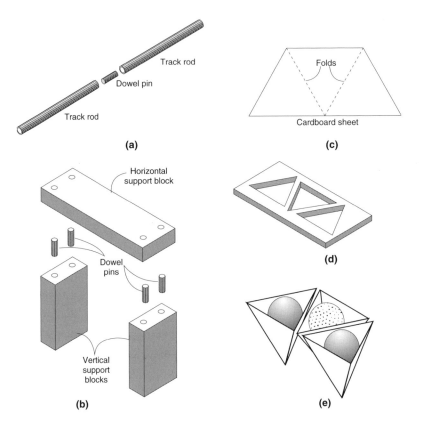

Figure 10.12 (a) Track rod sections are connected by dowel pins. (b) Track supports consist of one horizontal and two vertical blocks. (c) The cardboard sheet is folded into a trihedral cavity. (d) The cavities holder. (e) The arrangement of tetrahedral cavities when assembled in the holder.

stages of the run. In other words, the acceleration profile did not match the ideal case: the puller rod accelerated at the beginning only instead of continuously throughout the run.

PI: The speed of the puller rod is determined by the motor speed. The puller rod stops accelerating when the motor reaches its rated speed. The puller rod should be made to keep accelerating.

CS: A new custom pulley, shown in Fig. 10.14, was designed and built. The effective diameter of the pulley increases as the fishline is wound around it, resulting in ever-increasing tangential velocity, that is, continuous acceleration of the puller rod.

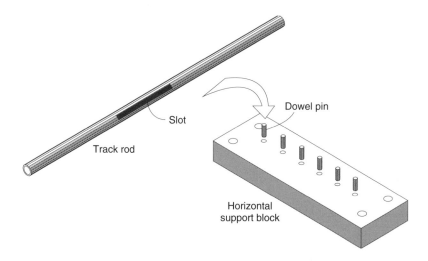

Figure 10.13 When assembled, the dowel pins prevent the transverse motion of the tracks.

Figure 10.14 Modified pulley geometry.

E: Further review of the videotape revealed that the initial acceleration was rather low due to the inertia of the moving components.

PI: Improve initial acceleration by increasing the initial torque.

CS: A small capacitor was added in parallel with the battery. Initially, the motor draws its energy from both the battery and the capacitor. The capacitor charges very quickly during the stopping motion of the balls.

E: The conceptual design was considered complete at this point, since the major issues had been addressed and initial tests showed that the total time was less than 1 second.

The overall assembly developed in this conceptual design process is shown in Fig. 10.15. Figure 10.16 is a summary of the parameter analysis process.

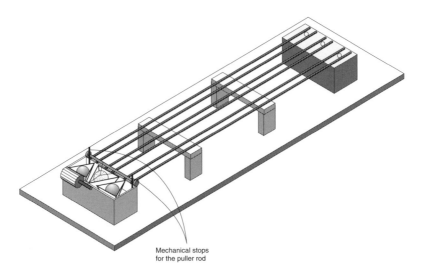

Mechanical stops
for the puller rod

Figure 10.15 Overview of the puller-rod system shown at the end of the run.

10.8 Discussion and Summary

Need analysis had a profound impact on the final design. The students systematically incorporated the characteristics of an ideal solution into their conceptual design using parameter analysis. The need analysis calculations provided the team with the ideal or theoretical scores. During the entire design process, they continuously checked the true performance against the ideal performance and strove to match the two. Continually evaluating the current score against the ideal one during the design process helped the team to assess their progress. The student designers entered the ASME Region VII Competition and won first place by transporting the balls over the full distance in under 1 second. Later they finished fourth at the ASME Winter Annual Meeting where thirteen ASME regional winners competed.

Let us look at one error committed by these students. In the kinetic energy calculation, they inadvertently ignored the angular velocity term. The balls roll when they are dragged from the platform, and they gain not only linear momentum but also angular momentum. Therefore, part of the energy output from the battery is wasted. An ideal solution should transport the balls linearly without spinning. This conceptual mistake probably cost the students first place. It could have been avoided by careful examination of

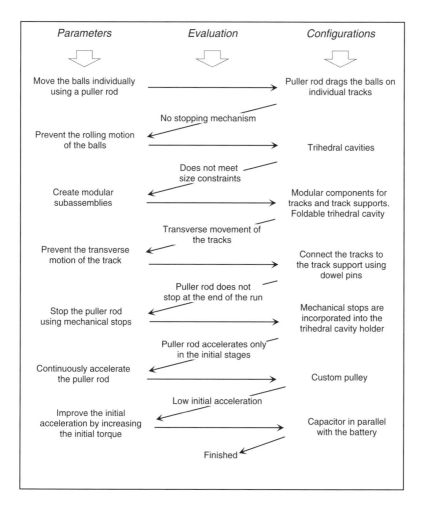

Figure 10.16 Summary of the parameter analysis process for conceptual design II.

implicit assumptions and development of more different conceptual designs to gain additional insights.

10.9 Thought Questions

1. Estimate the theoretical or ideal shortest time for moving the balls over the full 211.8-cm distance.

2. Estimate the percentage of energy lost due to the rolling motion of the balls.

3. Improve the design of the puller-rod ball mover by eliminating the spinning of the balls during transportation.

4. One possible initial idea for a ball mover, listed in the technology identification section, was a cable car. Develop this concept using parameter analysis.

5. Design a new ball mover device. Incorporate the new insights gained in this case study.

11

Technology Observation

This chapter proposes and demonstrates a particular approach to observing technology as a means of enhancing the conceptual design abilities of engineers. Designers can improve their skills by continuously understanding the underlying concepts in existing designs. For this process to be beneficial, however, it must be carried out at the appropriate level. The designer who does that will be able to develop a "bag of tricks" for future use.

11.1 Improving Design Abilities

Parameter analysis is not only a framework for carrying out innovative design, but also a method of developing skills in designers. In the conceptual design process, nothing can replace that spark of creativity which ignites good ideas. Some people are naturally blessed with more creative capabilities than others, but all designers need to practice the skill of conceptual design in order to develop it to its full potential. This is analogous to a naturally able athlete persevering through many hours of practice to attain a competitively high level of performance or to a famous actor who keeps attending drama classes throughout his professional life. In other words, a good designer must "practice" continuously to develop his or her skills. Every coach has a method by which he trains his athletes, and every drama teacher uses a particular technique of drama instruction. In a similar way, parameter analysis forms a coaching method that can sharpen the designer's inventive talents and ignite the spark of creativity more regularly.

The base on which the design methodology rests is an approach we simply call *technology observation* to indicate its purpose. This is not a specific step to be applied at a particular instance during design, but rather the continuous process of observing and analyzing existing technological products to study what others are doing. The important aspect of this observation is understanding *how* and *why* rather than merely *what* has been done. By observing technology in a particular way, over the years the designer will accumulate a knowledge base, or bag of tricks, that consists of understanding the underlying features and concepts of configurations and phenomena, as opposed to details of particular designs. When observed technology is applied to creating a new design, these underlying principles, in contrast to design specifics, will allow the designer to draw useful analogies and gain insight into the task at hand.

Technology observation is similar to "reverse engineering," but it needs to be done at a level commensurate with conceptual design by parameter analysis. This means that just as parameters—the concepts and underlying principles—drive the creation of configurations during design, so they also are the targets when the process is reversed, and existing configurations are observed. It is not enough to study products at the level of *what* they do, but rather to identify their key parameters by analyzing what is critical to their successful operation. In other words, we should try to gain the same insight into the design task of the observed product as the original designer had. An example of technology observation as applied to gas water heaters is presented next.

11.2 Domestic Gas Water Heaters Example

An inventor proposes a low-cost modification to present gas hot water heater configurations, which will give significant increases in the overall efficiency of operation. An increase on the order of 10% is claimed, with very strong emphasis on the low-cost specification and its implied simplicity. Let us perform technology observation on conventional gas water heaters and the modified version in order to demonstrate the level of understanding at which such process should take place.

Figure 11.1a is a schematic of a typical natural gas hot water heater. Combustion takes place in a burner at the bottom of the storage tank. Air is drawn in around the burner, and the combustion products, as well as excess air, flow up the stack. The stack begins as a tube passing through the middle of the tank. A typical unit will have a sheet metal insert in this tube which reaches down to about the beginning of the tubular section in the tank. The tank is surrounded by fiberglass insulation and a sheet metal cover over the insulation. Inlet and outlet pipes for cold and hot water are also shown on the schematic. Control of the burner is provided by a thermostatic system that acts to keep the water in the tank at a specified temperature.

The schematic in Fig. 11.1b describes the new design to be studied. The configuration created by the inventor consists of an expanded stack near the top of the water heater, filled with a large number of fins, each in intimate contact with the wall of the tank. These metal fins provide a large increase in the heat transfer surface in contact with the hot flue gases. Except for a slight decrease in the volume of water stored and the fact that the inlet and outlet pipes were moved during the design change, the new water heater configuration is quite similar to the present system.

In this example, the recognition of the primary feature of the design task, which the inventor is attempting to take advantage of, is relatively straightforward. By increasing the heat transfer surface by a very significant percentage, the inventor hopes that more of the heat that is contained in the stack gases will be transferred to the enlarged surface area of the stack, thereby increasing the efficiency of the water heater. The principal mechanism by which this is to be accomplished is, of course, convective heat transfer.

In order to gain a better understanding of the invention, we will investigate briefly the convective heat transfer that takes place in the unmodified water heater (Fig. 11.1a). For purposes of approximation, we will consider a water heater of the following size. The natu-

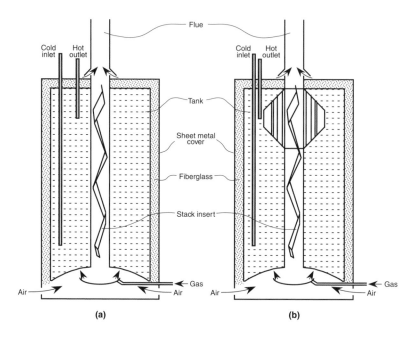

Figure 11.1 Schematics of (a) a conventional gas water heater and (b) the new design.

ral gas is flowing at a rate of 50,000 Btu equivalent heating value per hour, with the nominal efficiency of the water heater at 70%. The stack is about 4 feet high and 4 inches in diameter, and the tank holds about 100 gallons of water. We will assume that combustion is taking place with about 100% excess air.

The amount of heat transferred convectively is determined by the following equation:

$$\dot{Q} = h \, A \, \Delta T_{\mathrm{m}}$$

where $\dot{Q}$ is heat flow in Btu per hour, h is the heat transfer coefficient in Btu per square foot hour degree F, A is the surface area of the stack, and ΔT_{m} is the log mean temperature difference. ΔT_{m} can be found from

$$\Delta T_{\mathrm{m}} = \frac{\Delta T_1 - \Delta T_2}{\ln(\Delta T_1/\Delta T_2)}$$

where ΔT_1 is the temperature difference between the hottest part of the stack and its corresponding wall temperature, and ΔT_2 is the temperature difference between the top of the stack and its corresponding wall temperature. Using typical values for temperatures, which are about 2500°F at the bottom of the stack and 500°F at the top of the stack, we find that in order to transfer 35,000 Btu per hour to the water, we require that h take on a value of over 6 Btu per square foot hour degree F. Either from experience or a little investigation, we know that such a value for h is unrealistically high. However, to make sure of this assertion, we can estimate what h might be. The following equation is an approximate relationship for h as a function of the velocity of the gases and the stack diameter:

$$h = 0.43 \frac{V^{0.75}}{D^{0.25}}$$

V in this equation is the velocity of the free air volume—that is, the velocity if the stack gases were at standard pressure and temperature. The units for V are feet per second and for D, feet. Using this approximation, for a water heater sized as we assumed, we find that h is a little over 1.5. Thus, we find a very significant difference between the amount of heat transferred convectively and the total amount of heat transferred to the water.

A quick estimate tells us that there is enough energy in the stack gases leaving the water heater to more than yield the efficiency increase required, while keeping the flue gases above the condensation temperature of the water vapor produced during combustion in order to avoid corrosion within the system. In briefly investigating the principal mechanism of the new invention, we have been led to the conclusion that *the majority of the heat transferred to the water is done so by some other mechanism.* This may be a surprising result in light of the general shape of the water heater configuration and our normal intuition.

Three other mechanisms may account for this heat transfer. The first is radiation from solid parts near the burner; the second is radiation from the gases themselves; and the third is conduction. We know that the conduction phenomenon is small compared to other mechanisms in this case, and thus we are led to investigate radiative heat transfer. The following equation is an approximate relationship for the heat transfer per unit area due to radiation at temperature T_1 being absorbed at temperature T_2, where T_1 and T_2 are in degrees Rankine:

$$\frac{\dot{Q}}{A} = 0.15 \left[\left(\frac{T_1}{100}\right)^4 - \left(\frac{T_2}{100}\right)^4 \right]$$

A quick calculation shows us that a tremendous amount of heat can be transferred radiatively, particularly at the bottom of the tank, owing to the dependence of $\dot{Q}$ on the fourth power of temperature. (In fact, visual observation of the water heater in operation indicates that the principal function of the stack insert is to provide a sheet metal area that can glow at a high temperature and thus produce radiative heat transfer.) Finally, information on the amount of energy transferred radiatively by the combustion products can be obtained in handbooks. For our problem, this will be relatively constant regardless of configuration and so will not be pursued further here.

Quite obviously, the mechanisms of operation in a gas hot water heater are interrelated, thus producing a rather complex situation. During technology observation, we would like to gain further information on the relative magnitude of these interrelationships. Among the results pertinent to our example, we find the following:

1. An estimate of the area necessary for the additional convective heat transfer indicates that a nearly 200% increase in the total area is required. The temperature of the stack gases still remains above the condensation point.
2. An additional 1.5 inches of fiberglass insulation can yield an additional 5% in overall efficiency. Additions of even greater thickness of insulation, of course, yield diminishing returns. Any changes in the thickness of insulation require the modification of the outer shell of the hot water heater and must be evaluated carefully.

3. We recall that extremely sharp points in a flame glow at very high temperatures, and this phenomenon leads us to investigate further the possibilities of radiation enhancement as an alternative to the basic concept represented in Fig. 11.1b. The addition of radiating material into the burner area is a very minor modification, and thus such an alternative is a factor in determining the potential of this invention. For a wire to radiate at a particular rate, it must also receive heat at the same rate. Thus, there must be equilibrium between heat transferred convectively to the wire and that radiated by the wire. The following is an approximate relationship for the heat transferred convectively per unit area to a wire:

$$\frac{\dot{Q}}{A} = (T_s - T_1) \left[0.63 + 0.266 \frac{(T_s - 460)}{1000} \right] \frac{V^{0.56}}{D^{0.44}}$$

T_s is the temperature of the hot gas glowing over the wire, and T_1 is the temperature attained by the wire, both in degrees Rankine. The diameter D of the wire is in feet, and V is the free air velocity in feet per second. For the equilibrium conditions, the last two equations must yield the same value for $\dot{Q}/A$. The reader can quickly determine that wires on the order of a thousandth of a foot in diameter can be very effective additions near the burner in order to enhance radiation.

To complete the assessment of radiation enhancement, we need to remember that an increase in radiative heat transfer at the bottom of the tank will lower the temperature of the stack gases that begin to flow up the stack, thereby reducing the amount of heat transferred convectively. An approximate calculation of this interrelationship indicates that a 10% increase in radiative heat transfer will yield an overall increase of about 5.6%. Modifications required for radiation enhancement are minimal and will require little, if any, change to the basic configuration of the water heater.

The inventor has developed a low-cost, simple configuration aimed at improving the overall efficiency of a gas hot water heater. Based on the information generated and on our knowledge and experience, technology observation has shown that the invention will work, provided enough area is added without occluding the natural convection of the stack. However, adding sufficient area may be impractical. More importantly, our evaluation has identified other significant areas that also may lead to new configurations that could meet the requirements set forth.

11.3 Discussion and Summary

Although this example is far from a complete analysis of water heater operation, we have uncovered the essential elements to the extent that we could use the configuration or draw on our understanding of the phenomena in application to other situations. We can identify several concepts that we should put in our bag of tricks, or our mental knowledge base of building-block parameters.

The example illustrates several interesting features of the technology observation methodology. Most importantly, we have seen how an investigation of the operation of the water heater and an identification of the principal parameters involved have led us to some possibly unexpected conclusions. If our evaluation had just determined whether or not the new concept would work, we would have fallen short of uncovering significant information about radiative heat transfer in a hot water heater. Second, our study led us to an alternative scheme, the enhancement of radiative heat transfer, illustrating how technology observation can build up one's inventive skills. In fact, this understanding of radiative versus convective heat transfer becomes an innovation building block. Third, we identified several other parameters in the overall operation of the water heater which may be stored in our bag of tricks.

Very often the step of identifying concepts or parameters is not as straightforward as it is in this example, and significant mental effort may be required before the true focus of a technological product or invention is recognized. However, it is difficult to perform a good technology observation unless the principal parameters are truly understood.

Designers should actively and continually be engaged in technology observation to learn in this fashion and depth what the surrounding technology teaches. This will sharpen the designer's skill and thought process in identifying critical issues in all phases of the design process. Moreover, such systematically cataloged knowledge, in the form of general principles and concepts, becomes invaluable for later use in conceptual design.

11.4 Thought Questions

The foregoing approach to understanding the critical concepts behind the output of other innovators should be a life-long activity. The reader is encouraged to practice this approach on a regular basis as a means of continuously adding to his or her expertise.

11.5 Bibliography

Information about gas water heaters that may be useful to understanding the example can be found in:

Avallone, E. A. and Baumeister, T. *Marks' Standard Handbook for Mechanical Engineers.* 10th ed. New York: McGraw-Hill, 1996.

Incropera, F. P. and DeWitt, D. P. *Fundamentals of Heat and Mass Transfer.* 4th ed. New York: John Wiley & Sons, 1996.

12

Conclusion

This chapter summarizes the approach of this book to the initial stages of the design process, namely, need identification and analysis and conceptual design. Cognitive aspects of parameter analysis are discussed next, followed by a description of its relationships to other design methodologies. The chapter concludes with a brief overview of how the design process should proceed through concept selection, embodiment design, and detail design.

12.1 The Essence of the Methodology

The ultimate goal of the approach to conceptual design presented in this book is to help designers to "think better." In this context thinking better means being more creative and innovative, and doing it faster. In today's competitive market, a successful product needs to perform better than its rivals and must also be introduced earlier.

The key elements of the early design stages introduced throughout the book are:

- *Need Identification:* Clarification of what should and what should not be designed. The main obstacle to avoid here is following customers' configurational task statements.
- *Need Analysis:* The thorough study of everything that may be related to the design task. We recommend organizing this

stage according to the following five general categories: *performance, value, size, safety,* and *special.*

- **Design Requirements:** The functions and constraints generated during need analysis are summarized as a set of concise and quantitative specifications.
- **Technology Identification:** Conceptual design begins by naming as many general approaches for addressing the task as possible. These core technologies are assessed for their chances of success and serve as starting points for concept development.
- **Parameter Analysis:** Conceptual design is carried out by continuously cycling through *PI–CS–E* (*parameter identification–creative synthesis–evaluation*) triplets. Every configurational step must be conceptually backed by a *parameter*—idea, physical insight, relationship, justification, and so on—and evaluated critically, yet constructively. This iterative process continues until all the design requirements have been satisfied and no unresolved issues remain.

12.2 Cognitive Aspects of Parameter Analysis

As we design, we spend time and resources to make progress toward creating a solution. The key to successful design is not just making progress but *making progress at the desired rate.* Thus, the designer is faced with the challenge of controlling the design process to achieve fast progress. As with any process, we must understand the process in order to control it. In design, if we know:

- *How we design when faced with a design task,* and
- *What the pitfalls of the natural thought process are,*

then we can:

- *Be conscious of our thought process.*
- *Identify when progress is not being made at the desired rate.*
- *Take corrective actions to become more effective.*

Let us try to understand the natural thought process—that is, the course followed by humans when faced with a design task—and some of the potential pitfalls. We can divide this thought process into two distinct stages:

1. Recognition of the design task.
2. Development of the design solution.

Recognition of the Design Task

Humans recognize objects by their resemblance to the prototypes representing various knowledge categories. For instance, we identify a robin as a bird because it closely resembles the prototype of the bird category. The prototypes closely resemble the typical examples in their category. A bird prototype resembles a robin more closely than a penguin. This fact is reflected in the increased time to recognize penguins as birds. A similar recognition process occurs when a designer is presented with a design task.

During the initial stages, the designer categorizes the task based on the prototypes and not on concepts or scientific principles. The recognition process is very fast and may occur even after reading only a small portion of the problem statement. It is based on the configurational issues, since prototypes are configurational in nature. At the end of the recognition process, the designer would therefore have activated a prototype that could be either concrete or abstract depending on the scope of the problem.

Development of the Design Solution

The development process involves modifying, adapting, or sizing the prototype to satisfy the new conditions, constraints, and requirements of the design task. A phenomenon known as *design fixation* plays an important role in the development stage. It prevents the designer from considering alternative solutions.

In most cases, the core concept behind the prototype remains unchanged during the development process because of the difficulty in accessing the concepts behind the prototype. For instance, for most people the prototype for manholes is circular and that of airplane wings is swept backward. However, the fundamental concepts that dictate the geometry may neither be easily accessible nor existent in the designer's mind. Thus, the designer changes small configurational details without gaining deeper understanding.

Pitfalls of the Natural Thought Process

Let us look at a typical thought process during the design of a toy for blind children. The design task automatically invokes the prototype for the toy category. This prototype may be different for various people depending on their background and gender. The designer then modifies the toy to satisfy the needs of blind children. It is usual for designers who do not follow a systematic methodology to ignore issues such as gender, age, sensory mechanisms other than touch/feel, size of the toy, ability to find the toy, and color. This example illustrates some of the key pitfalls of the natural thought process:

1. The designer develops the solution without proper understanding of customer needs. Such activity results in inferior products that do not appeal to the customers.
2. Due to *design fixation*, designers do not consider alternative solutions and develop the prototype into the final solution. Since the prototype resembles the typical examples in the category, the final solution is most likely not innovative.
3. The attention is on configurational issues and details. Key issues emerge as late as during the testing phase and result in long product realization times.
4. The development activity revolves around sizing, configurational modification, and adaptation. There is no real synthesis to create the solution.
5. There is a natural tendency to drift from the conceptual domain to the configurational domain. In contrast, the movement from the configurational to the conceptual domain requires considerable cognitive effort. As a result, the designer may ignore critical problem areas.

The systematic methodologies of this book, namely, need identification and analysis and conceptual design by parameter analysis, attempt to overcome the potential pitfalls of the natural thought process, as summarized in Table 12.1.

Table 12.1 How the methodologies of this book overcome pitfalls of the natural thought process.

Potential Pitfalls of the Natural Thought Process	Benefits of the Systematic Methodologies of This Book
Natural tendency to design without proper understanding of the design task.	The *need identification and analysis* methodology suppresses this tendency and defines the need in solution-neutral terms.
Due to design fixation, only the prototype is developed into the final solution. The solution is not innovative since it closely resembles the typical examples.	The *technology identification* step provides a breadth of ideas and several starting points. It reduces the design fixation and thereby increases the possibility of creating an innovative solution.
Focus is on the detailed issues.	*Parameter identification (PI)* focuses attention on the key conceptual issues in the design.
Development involves sizing, configurational modifications, and adaptations.	*Creative synthesis (CS)* increases innovative content and establishes synergy between basic components of the design.
Movement from configuration to concept space is difficult.	*Evaluation (E)* provides an in-depth understanding and aids in moving from configuration to concept space.

12.3 Relation to Other Design Methodologies

Several existing design methods and practices have parameter analysis "flavor." Suh's *axiomatic design* (see Suh, 1990) states a model of the design process wherein mapping takes place from the functional space to the physical space. The functional space consists of functional requirements (FRs), and the physical space contains design parameters (DPs). Note that while this terminology is very different from what we use in parameter analysis (we use concept and configuration spaces, and the elements in concept space are the ones dubbed *parameters*), the essence is similar. Suh also presents design as a sequential process, in which the most important functional requirements should be identified first, a physical solution created, leading to new, lower-level functional requirements, and so on.

The principles that the mapping technique must satisfy to produce a good design are the following two axioms:

1. Maintain the independence of functional requirements.
2. Minimize the information content of the design.

Many corollaries can be derived from these two axioms to offer the designer assistance in synthesizing physical solutions to satisfy the functional requirements. The similarity between axiomatic design and parameter analysis implies that the two methodologies could perhaps be combined.

Dixon and Poli (1995) use a design methodology dubbed *guided iteration*, which consists of four general steps: formulating the problem, generating alternative solutions, evaluating the alternatives, and if needed, redesigning. This four-step process is applied at the conceptual design level, as well as at the later stages of configuration and parametric design. (These correspond to our embodiment and detail design, respectively.) Two approaches are offered for generating the conceptual design alternatives. The first is based on functional decomposition, and the second calls for searching for suitable physical laws and effects. The latter emphasis on underlying principles is reminiscent of our *parameters*. The last step in the methodology, redesign, is guided by the results of previous evaluations, by qualitative physical reasoning, and by knowledge of manufacturing processes. As such, it too resembles the iterative nature of parameter analysis.

French (1999) presents numerous thoughtful examples of conceptual design principles and their application to different problems. Many of those so-called synthesizing principles and design rationales can be shown to be similar to what this book calls *design parameters*. Also resembling parameter analysis are French's listed objectives of the design methods, which are:

1. Increasing insight into design tasks and acquiring the insight fast.
2. Diversifying the approach to the design task.
3. Reducing the size of the mental steps during design.

4. Prompting inventive steps and reducing the chances of over-looking them.
5. Generating problem-specific design philosophies.

Several other design books, mentioned in the bibliography at the end of this chapter, emphasize the systematization of the design process. Such systematization is different from the approach of this book, which attempts to "unlock" the cognitive processes that take place in successful conceptual design.

12.4 What's Next?

This book purposefully focuses on that portion of the design process that begins with an initial need and ends with conceptual design. Clearly, much work is still needed to turn a conceptual design into a product. The downstream stages are outlined below.

Concept Evaluation and Selection

An important characteristic of parameter analysis—the conceptual design methodology of this book—is that evaluation is an inseparable component of the concept development process. This means that by the time conceptual design is finished, many evaluation steps have been completed. As a result, the conceptual design has already been judged to meet all the design requirements and to constitute a feasible solution that is based on ready technology. Common concept evaluation techniques, which compare the design with the requirements, study its feasibility, or assess its technology readiness, are therefore all unnecessary when conceptual design is done by parameter analysis. However, these approaches may con-tribute to the evaluation steps along the way.

The conceptual design stage often involves developing several alternative designs. This is a common practice when a completely new product is designed. Several designers or design teams are assigned the same task, each designer or team performs an inde-pendent conceptual design, and the results are evaluated during a design review. Realistic constraints on development cost, time, and resources usually preclude the possibility of finalizing the alternative designs. Instead, the choice among the competing designs needs to be made earlier, at the end of the conceptual design stage.

Concept selection is not an easy task. The main difficulty lies in basing the evaluation and decisions on order-of-magnitude information associated with conceptual designs, as opposed to the complete quantitative data available for finished designs. This may be even more challenging in situations where the conceptual design stage terminates after listing only general ideas and solution approaches. Fortunately, parameter analysis requires that all the major issues and aspects of a design be addressed, thus resulting in a relatively detailed description of the design. Not only are the designs quite well developed at the end of our conceptual design stage, but the development process itself is so elaborate that much insight into the task and the solution has been gained by this time.

The most suitable selection method to apply to conceptual designs generated by parameter analysis is Pugh's method, which is based on a comparison matrix (Pugh, 1991). Concepts are listed as columns, and evaluation criteria as rows. One concept is chosen as datum, and every other concept is assigned a "+" if it is better, "−" if it is worse, and an "S" if it is the same as the datum, for each criterion. Next, the pluses and minuses are summed, and the winning concept is selected. There are some variations on the basic theme, such as scoring on a numerical scale with and without relative weights, but we shall not describe this method in detail because it is well documented in many books.

However, a word is in order regarding the choice of the evaluation criteria. The design requirements are usually not suitable as criteria because all the concepts that reach this stage have been shown to satisfy the requirements. This means that a set of criteria needs to be developed especially for selecting the concept. With the added insight gained during conceptual design, and keeping in mind the original need, criteria such as "the ease with which property X can be achieved" and "the robustness of the design" should be used. In any case, the designer should apply Pugh's method for concept selection as a tool and not let the method make the decision for him or her. It is recommended that the total number of pluses, minuses, or the difference between them not be used blindly. Instead, the designer should use the comparison matrix to examine the specific strengths and weaknesses of the concepts.

Embodiment Design

Having selected the most promising concept for further development, the design process progresses to the *embodiment design* stage. This stage is also known as *preliminary design, layout design,* or *configuration design.* When conceptual design ends with vague and general ideas, embodiment design becomes a major development stage. However, with the detailed conceptual designs that result from parameter analysis, not that much is left to do here.

The major task during embodiment design is firming up the product structure. Separate components and subassemblies need to be defined now, and a layout drawing should be produced. Issues that were judged noncritical during conceptual design should be addressed at this stage. For example, a specific electric motor required by the design, the type of gears to use for speed reduction, or the class of material (e.g., steel, plastic) can be specified. Preliminary calculations may also be required to size components.

The conceptual design methodology of parameter analysis may also be applied to embodiment design. The philosophy of handling one design aspect at a time, focusing on the more important issues first, and supporting configurational decisions with concepts is still useful. Principles commonly used in embodiment design, such as "use short and direct transmission path for forces," "match manufacturing process to production volume," and "reduce part count to improve ease of assembly," are all concepts or parameters that affect the configuration.

Detail Design

Comprehensive calculations of strength and other properties, precise dimensioning and tolerancing, material specification, and many other details are produced during this final stage of design. The remainder of the product realization process, which may involve prototyping, testing, and the like, is beyond the scope of this book.

12.5 Bibliography

Some of the cognitive aspects of parameter analysis are studied in the following articles:

Jansson, D. G., Condoor, S. S., and Brock, H. R. "Cognition in Design: Viewing the Hidden Side of the Design Process." *Environment and Planning B: Planning and Design* 19 (1992): 257–271.

Jansson, D. G. and Smith, S. M. "Design Fixation." *Design Studies* 12, No. 1 (1991): 3–11.

Axiomatic design is the subject of:

Suh, N. P. *The Principles of Design.* New York: Oxford University Press, 1990.

The *guided iteration* design methodology is used in the following textbook:

Dixon, J. R. and Poli, C. *Engineering Design and Design for Manufacturing: A Structured Approach.* Conway, MA: Field Stone Publishers, 1995.

Many examples with parameter analysis "flavor" can be found in:

French, M. *Conceptual Design for Engineers.* 3rd ed. London: Springer-Verlag, 1999.

Some books that emphasize systematic design processes are:

Cross, N. *Engineering Design Methods.* Chichester: John Wiley & Sons, 1989.

Pahl, G. and Beitz, W. *Engineering Design: A Systematic Approach.* London: The Design Council, 1988.

Stoll, H. W. *Product Design Methods and Practices.* New York: Marcel Dekker, 1999.

Ullman, D. G. *The Mechanical Design Process.* 2nd ed. New York: McGraw-Hill, 1997.

Ulrich, K. T. and Eppinger, S. D. *Product Design and Development.* 2nd ed. New York: McGraw-Hill, 2000.

Our recommended procedure for *concept selection* is described in:

Pugh, S. *Total Design.* Wokingham, England: Addison-Wesley, 1991.

Index